From Out of the
Shadows

From Out of the Shadows

MEXICAN WOMEN IN

TWENTIETH-CENTURY

AMERICA

Vicki L. Ruiz

New York   Oxford
OXFORD UNIVERSITY PRESS
1998

Oxford University Press

Oxford   New York
Athens   Auckland   Bangkok   Bogotá   Bombay
Buenos Aires   Calcutta   Cape Town   Dar es Salaam
Delhi   Florence   Hong Kong   Istanbul   Karachi
Kuala Lumpur   Madras   Madrid   Melbourne
Mexico City   Nairobi   Paris   Singapore
Taipei   Tokyo   Toronto   Warsaw

and associated companies in
Berlin   Ibadan

Published by Oxford University Press
198 Madison Avenue, New York, New York 10016

Oxford is a registered trademark of Oxford University Press

Library of Congress Cataloging-in-Publication Data
Ruiz, Vicki.
From out of the shadows: Mexican women
in twentieth-century America / Vicki L. Ruiz.
p.   cm.   Includes bibliographical references and index.
ISBN 0-19-511483-3
1. Mexican American women—History—20th century.
2. Mexican American women—Social conditions.
I. Title   E184.M5R86   1997
305.48'86872073—dc21   97-9387

Permission credits:
Sections of Chapter 2 were published as
"Dead Ends or Gold Mines: Using Missionary Records
in Mexican American Women's History,"
Frontiers 12:1 (1991): 35–56.

An earlier draft of Chapter 3 was published as
"The Flapper and the Chaperone: Historical Memory Among Mexican American Women"
in *Seeking Common Ground: A Multidisciplinary Reader
on Immigrant Women in the United States,*" ed. Donna Gabaccia (Westport, CT:
Greenwood Press, 1992), pp. 141–157.

"University Avenue" by Pat Mora
is reprinted with permission from the publisher
of *Borders* (Houston: Arte Público Press, 1986).

1 3 5 7 9 8 6 4 2

Printed in the United States of America
on acid-free paper

FOR THE STORYTELLERS WHO GAVE ME HISTORY LESSONS AT HOME

In memory of my grandmother
María de las Nieves Moya de Ruiz
(1880–1971)

and to my mother
Erminia Pablita Ruiz Mercer
(1921–)

Contents

◆✖◆

Acknowledgments   *ix*

Introduction   *xiii*

1. Border Journeys   *3*

2. Confronting "America"   *33*

3. The Flapper and the Chaperone   *51*

4. With Pickets, Baskets, and Ballots   *72*

5. La Nueva Chicana: Women and the Movement   *99*

6. Claiming Public Space   *127*

   Epilogue   *147*

   Appendix   *153*

   Notes   *159*

   Bibliography   *209*

   Index   *231*

Acknowledgments

❖

THIS book would not be possible without the voices of the individuals who have shaped this narrative, as historical actors, as scholars, and as friends. First, I would like to thank the following people who shared with me their memories: Eusebia Buriel, Ray Buriel, Elsa Chávez, Carmen Bernal Escobar, Alma Araiza García, Fernando García, Martha González, Dorothy Ray Healey, Lucy Lucero, Ernest Moreno, Graciela Martínez Moreno, the late Luisa Moreno, Julia Luna Mount, Concha Ortiz y Pino de Kleven, María del Carmen Romero, my mother Erminia Pablita Ruiz Mercer, and Jesusita Torres. I thank my former students Carolyn Arredondo, Jesusita Ponce, Lydia Linares Peake, and David Ybarra for giving me permission to cite their oral interviews. With generosity and encouragement, Sherna Berger Gluck has allowed me to quote from several volumes of the *Rosie the Riveter Revisited* oral history collection housed at California State University, Long Beach.

During my ten years of wandering in and out of archives, staff members have been extraordinarily helpful and I express a deep appreciation to Christine Marín, Special Collections, Arizona State University; Rose Diaz and Tom Jaehn, Special Collections, Zimmerman Library, University of New Mexico; Kathcrine Kane and Anne Bonds, Colorado Historical Society; Kate Adams, Barker History Center and Margo Gutiérrez, Benson Library, University of Texas, Austin; and María E. Flores, Our Lady of the Lake College,

San Antonio. I also thank the helpful staff at the Bancroft Library, University of California, Berkeley; Special Collections, University of California, Los Angeles; Southern California Library for Social Studies and Research (especially Sarah Cooper); Colección Tloque Nahuaque, Chicano Studies Library, University of California, Santa Barbara; Western History Department, Denver Public Library; Houchen Community Center, El Paso; Daughters of the Republic of Texas Library at the Alamo; Labor Archives, University of Texas, Arlington; DeGroyler Library, Southern Methodist University; University Archives, New Mexico State University; and the Arizona Historical Society Library, Tucson.

A debt of gratitude goes out to Victor Becerra, Ernie Chávez, Tom Jaehn, Valerie Matsumoto, Lara Medina, Beatríz Pesquera, Denise Segura, Howard Shorr, and Devra Weber for taking time out of their own busy schedules to share with me photographs, newspaper clippings, and primary materials. Thank you more than words can convey. I also acknowledge friends and colleagues who sent me their published and unpublished papers, works that have unquestionably enriched this manuscript: Gabriela Arredondo, Ray Buriel, Gilbert Cadena, Roberto Calderón, Angel Cervantes, Ernie Chávez, Marisela Chávez, Kenton Clymer, Virginia Espino, Jeff Garcílazo, Ramón Gutiérrez, Tom Jaehn, Anne Larson, Margo McBane, Jesus Malaret, Valerie Matsumoto, Lara Medina, Cynthia Orozco, Naomi Quiñonez, Ada Sosa Riddell, Margaret Rose, George Sánchez, Marjorie Sánchez-Walker, Denise Segura, María Soldatenko, Richard Street, Quintard Taylor, and Zaragosa Vargas.

I have been blessed with the privilege of working with a number of highly motivated and talented graduate students, people who will definitely make a difference in our profession. The UC Davis cohort includes James Brooks, Kevin Leonard, Jennifer Levine, Matthew Lasar, Jesus Franscisco Malaret and of course, the "Sisterhood"—Kathleen Cairns, Annette Reed Crum, Margaret Jacobs, Olivia Martínez-Krippner, Alicia Rodríquez-Estrada, and Yolanda Calderón Wallace. The Claremont crew wants to set the world on fire: Frank Barajas, Virginia Espino, Matthew García, Timothy Hodgdon, Alice Hom, Peg Lamphier, Matthew Lasar, Lara Medina, Marian Perales, Naomi Quiñonez, Alicia Rodríquez-Estrada, Arlene Sánchez-Walsh, Emilie Stoltzfus, Mary Ann Villarreal, and Antonia Villaseñor are comitted to the bridging of the academy and the community. I have also enjoyed my many conversations with Pat Ash, Philip Castruita, Antonia García, Lee Ann Meyer, Kat Norman, and Amanda Pérez. A special thanks is reserved for

Marisela Chávez, Virginia Espino, Timothy Hodgdon, Peg Lamphier, and Mary Ann Villarreal, my current graduate students who migrated with me to the "dry heat" of Arizona and to the daughters of the desert, Rose Diaz, Christine Marín, and Jean Reynolds, who have made us all feel so welcome.

Institutional support for this manuscript has come from many sources. An American Council for Learned Societies Fellowship proved crucial in the early phases of this project. A UC Davis Humanities Fellowship and a faculty development award allowed me to take a one-year sabbatical from the classroom. In addition, research funds associated with the Mellon Humanities chair that I held for three years at The Claremont Graduate School along with a summer research grant and a Haynes Fellowship provided funds to complete the archival research and transcribe interviews. I wrote the last chapters at Arizona State University where I recieved a one semester sabbatical.

At UC Davis, my undergraduate assistants (both are now attorneys) Amagda Pérez and Viola Romero were conscientious to a fault. Thanks also to Ada Arensdorf and Eve Carr. The meticulous efforts of Timothy Hodgdon and Matthew Lasar proved invaluable in preparing the manuscript for publication. Mis compañeras/os Angie Chabram-Dernesesian, Ed Escobar, Estelle Freedman, Gayle Gullett, Gail Martínez, Valerie Matsumoto, Beatríz Pesquera, Mary Rothschild, Raquel Salgado Scherr, Howard Shorr, Susan Tiano, and Clarence Walker helped me regain my focus during a very difficult time in my life.

At various stages, several individuals offered inspiration and constructive criticism. I would like to thank Ramón Gutiérrez, George Lipsitz, Elizabeth Martínez, Valerie Matsumoto, and Howard Shorr for their careful readings of one or more chapters. The ASU Women's History Reading Group, particularly Michelle Curran, Susan Gray, Gayle Gullett, Mary Melcher, and Sybil Thornton, provided support and insight. Peggy Pascoe and Sarah Deutsch read the entire manuscript and their comments (and Peggy's line editing) substantially strengthened the narrative as a whole. Peggy y Sally—gracias por todo.

I have felt privileged to work with Sheldon Meyer, a steadfast advocate and extraordinary editor. I also acknowledge the careful attention paid to this manuscript by Brandon Trissler and Helen Mules. Their professionalism and enthusiasm eased this overprotective author. Rosemary Wellner, too, deserves thanks for her skills as a copy editor.

To my knowledge, there are only fourteen Chicanas with PhDs in history. I am number four. Often we labor alone, subject to "proving" our research and our very presence in the academy. I would like to acknowledge the labors of Louise Kerr, Antonia Castañeda, Deena González, Camille Guerin Gonzáles, Gloria Miranda, María Montoya, Lorena Oropeza, Cynthia Orozco, Emma Pérez, Naomi Quiñonez, Yolanda Romero, Elizabeth Salas, and Shirlene Soto and to honor the legacies of the late Irene Ledesma and Magdalena Mora.

There are no words to describe the appreciation and love I have for the wonderful men who have graced my life. When I completed *Cannery Women, Cannery Lives*, my sons Miguel and Danny were six and three. In ten years, Miguel has gone from decorating the door with crayons to enjoying new challenges with a driver's license and Danny's taste in television has changed from *Sesame Street* to MTV. I value their patience and resilience in the midst of our several moves. They have borne their absentminded mother with good humor and much love. My father Robert Mercer passed away into the next life on November 14, 1995, after a long illness, but I feel his presence and independent spirit. A man of honor, gentle strength, with a great capacity for love, Victor Becerra, my husband, changed my life. Victor has contributed in many ways to the shaping of this work. Listening to scattered passages, reading preliminary drafts, and offering well-placed suggestions, he reminds me to write from the heart.

V. L. R.

Introduction

❖

W HEN I was a child, I learned two types of history—the one at home and the one at school. My mother and grandmother would regale me with tales of their Colorado girlhoods, stories of village life, coal mines, strikes, discrimination, and family lore. At school, scattered references were made to Coronado, Ponce De León, the Alamo, and Pancho Villa. That was the extent of Latino history. Bridging the memories told at the table with printed historical narratives fueled my decision to become a historian.

From Out of the Shadows focuses on the claiming of personal and public spaces across generations. As farm workers, flappers, labor activists, barrio volunteers, civic leaders, and feminists, Mexican women have made history. Their stories, however, have remained in the shadows.

The introduction to my first book, *Cannery Women, Cannery Lives,* refers to the shadowing of Mexican women's experiences. "Scholarly publications on Mexican American history have usually relegated women to landscape roles. The reader has a vague awareness of the presence of women, but only as scenery, not as actors . . . and even their celebrated maternal roles are sketched in muted shades." Little did I realize that this theme had also resonated among earlier chroniclers of Spanish New Mexico, most notably Cleofas Jaramillo. In 1941, she compiled a collection of folklore, *Sombras del pasado/Shadows of the Past,* in which she drew on the collective memories of her Hispano narrators.[1] Similarly, I have drawn on the collective voices of the women who have shaped and given meaning to my work.

Race, class, and gender have become familiar watchwords, maybe even a mantra, for social historians, but few get beneath the surface to explore their intersections in a manner that sheds light on power and powerlessness, boundaries and voice, hegemony and agency. This book addresses issues of interpreting voice and locating power between and within communities, families, and individuals. Women's lives, dreams, and decisions take center stage.

Women of Mexican birth or descent refer to themselves by many names—Mexicana, Mexican American, and Chicana (to name just three). Self-identification speaks volumes about regional, generational, and even political orientations. The term Mexicana typically refers to immigrant women, with Mexican American signifying U.S. birth. Chicana reflects a political consciousness borne of the Chicano Student Movement, often a generational marker for those of us coming of age during the 1960s and 1970s. Chicana/o has also been embraced by our elders and our children who share in the political ideals of the movement. Some prefer regional identification, such as Tejana (Texan) or Hispana (New Mexican). Spanish American is also popular in New Mexico and Colorado. Latina emphasizes a common bond with all women of Latin American origin in the United States, a politicized Pan American identity. Even racial location can be discerned by whether one favors an Iberian connection (Hispanic) or an indigenous past (Mestiza or Xicana).

As part of her stand-up routine, lesbian writer and comic Monica Palacios articulates her multiple identities as follows:

> When I was born
> I was of Mexican-American persuasion
> Then I became Chicana
> Then I was Latina
> Then I was Hispanic
> Then I was a Third World member
> (my mom loved that)
> Then I was a woman of color
> Now I'm just an Amway dealer
> And my life is happening.[2]

Literary critic Alicia Arrizón refers to Palacios's work as "one of challenge where humor becomes the tool of reconstructing ways of understanding the self." Poet and novelist Alicia Gaspar de Alba conveys the image of the Chicana writer as "the curandera (medicine woman) or the bruja (witch) . . . the keeper of culture, the keeper of memories."[3] The exploration of identities, the conservation and creation of cultural practices and traditions, and the reconstruction of historical narratives are not without political intent. In the words of Sonia Saldívar-Hull, "The Chicana feminist looks to her history . . . to learn how to transform the present."[4]

Focusing on the twentieth century and the Southwest, this book surveys women's border journeys not solely in terms of travel, but of internal migration—creating, accommodating, resisting, and transforming the physical and psychological environs of their "new" lives in the United States. These are journeys of survival, resiliency, and community. They reveal, to quote Chandra Talpade Mohanty, "the material politics of everyday life, especially the daily life struggles of poor people—those written out of history."[5]

In 1900, Mexican women had a long history of settlement in what is now the Southwest extending back to the Coronado expedition of 1540. Historians, including Ramón Gutiérrez, Albert Hurtado, Antonia Castañeda, Angelina Veyna, and James Brooks, have reconstructed the ideological beliefs and physical realities of women in the Spanish Borderlands across racial and social locations from the elite gentry to indentured servants.[6] In addition, Deena González, Sarah Deutsch, and Lisbeth Haas carefully delineate the lives of Spanish-speaking women in the decades following the U.S.–Mexican War (1846–1848), and, in placing women at the center of Hispano communities, they document gendered strategies for resisting political, economic, and cultural conquests.[7] Women's kin and friend networks—their comadres—were indispensable for both personal and cultural survival. Comadres helping comadres, neighbors joining neighbors—such patterns of mutual assistance run through the histories of Mexican–American women.

Through mutual assistance and collective action, Mexican women have sought to exercise control over their lives at home, work, and neighborhood. *From Out of the Shadows* opens with "Border Journeys," a chapter that details the ways in which Mexicanas claimed places, albeit economically precarious ones, for themselves and their families in the United States while "Confronting America" records their efforts to create their own cultural spaces. "The Flapper and the Chaperone" focuses on the second generation taking the first individual steps toward sexual autonomy. "With Pickets, Baskets, and Ballots" surveys women's activist paths in labor unions, voluntary associations, and political affiliations.

"La Nueva Chicana" examines the development of a distinctive Chicana feminist consciousness, a consciousness with historical antecedents predating the 1960s and with contemporary lessons for bridging individual and community empowerment. Covering the period 1970 to the present, "Claiming Public Space" emphasizes the ways in which women from the shop floor to city hall have made a difference in their lives and their neighbors' lives through community-based organizations.

It is important to situate this thread of public and private spaces that appears in each chapter. One's positionality inside the home, the community, and the workplace cannot be separated into neat categories of analysis. The feminist edifice of separate spheres need not apply as "the inextricable nature of family life and wage work in the histories of immigrant wives and women of color explodes the false oppositions at the heart of the public/private dichotomy."[8] Integration, rather than separation, provides a more illuminating construct in exploring the dynamics of Mexicana/Chicana work and family roles.

A second thread running through the narrative is that of cultural coalescence. Immigrants and their children pick, borrow, retain, and create distinctive cultural forms. There is not a single hermetic Mexican or Mexican–American culture, but rather permeable *cultures* rooted in generation, gender, and region, class, and personal experience. People navigate across cultural boundaries and consciously make decisions with regard to the production of culture. However, bear in mind that people of color have not had unlimited choice. Race and gender prejudice and discrimination with their accompanying social, political, and economic segmentation have constrained aspirations, expectations, and decision-making.

Though filtered through the lens of time and mediated by the interviewer, oral histories shed much light on individual stories of resistance, resilience, and creativity. It is not a question of "giving" voice, but of providing the space for people to express their thoughts and feelings in their words and on their own terms. Reclaiming, contextualizing, and interpreting their memories remain the historians' tasks. I am reminded of a line in William Blake's "Auguries of Innocence"—"To see a world in a grain of sand."[9]

In September 1992, the *Los Angeles Times* carried an article about a Mexican–American photo exhibit at the El Monte Historical Museum.[10] Though a bit timid behind the wheel, I gathered my courage and ventured onto the freeway bound for El Monte. Inside the museum, I lingered over pictures documenting migrant life: the Hick's Camp barrio, family weddings, and neighborhood celebrations. I even noticed images of the popular Long Beach nightspot, the Cinderella Ballroom, where couples posed in front of paper moons and caboose facades. An elderly Mexicana entered the room accompanied by her daughter. She began to identify all the people and landmarks captured in the photographs. I introduced myself and tagged along as she graciously gave us a history lesson. A few months later, Jesusita Torres invited me into her home and shared with me her remarkable story of survival and hope. Her journey unfolds here.

1

Border Journeys

❖

THE year is 1923; the place Gómez Palacios in the Mexican state of Durango. As she watched her mother pack a few belongings, Jesusita Torres was warned by her mother Pasquala Esparza not to tell anyone of their plans. Several days later, shortly before noon, Pasquala would sneak out of the family home with nine-year-old Jesusita at her side and one-month-old Raquel in her arms. They headed for the train station with Pasquala surveying the landscape for any signs of her husband or his relations. She must have breathed a sigh of relief as the train began its journey to Ciudad Juárez from which she hoped to cross with her children into the United States.

During our interview seventy years later, Señora Torres would reveal that her stepfather (Raquel's biological father) in Gómez Palacios had been cruel to her and her mother. In her words:

> I never knew my father. . . . My mother got married again and things did not work. I guess they did not work because I was mistreated, too, you know. So I think the only way she could get away was to come over here.[1]

Pasquala intended to stay in Juárez until she had the money to secure passports for herself and her children. Her destination was El Monte, California, to live with her married sister.

3

That same year in the village of San Julián in the Los Altos region of Jalisco, Petra Sánchez made plans to return to the United States with her husband Ramón, their three infants (two-year-old Guadalupe, one-year-old Librado, and newborn Margarita) and five-year-old José, the son of Ramón and his late wife Guadalupe Rocha. The villagers of San Julián may have thought Petra and Ramón an unusual couple. Theirs had not been a conventional courtship. Ramón's late wife Guadalupe had been Petra's older sister. When Guadalupe and her daughter Ampelia succumbed to the global influenza outbreak of 1918, Ramón decided that he and José should live with the Rochas. Whether by choice or arrangement, Ramón and Petra married in 1920 and then the newlyweds journeyed to California. Laboring as berry pickers in the vicinity of Buena Park, they hoped to make enough to return to San Julián with a nice nest egg. They arrived back to the village in 1923, but within a year the couple decided to make California their home.[2] Their second migration was marked by tragedy. When Ramón moved ahead to Buena Park leaving his family temporarily behind with his brother, baby Margarita died and, as Librado would later recall to his niece Marjorie Sánchez-Walker, "mother was alone . . . when Margarita died." By 1924, Ramón, Petra, and their increasing family worked in the fields of Knott's Berry Farm. About fifty miles to the north in El Monte, Pasquala and her daughter Jesusita would also be picking berries.[3]

Jesusita Torres and Petra Sánchez were part of the first modern wave of Mexican immigration to the United States. The society they entered was one already marked by multiple conquests, migrations, and overlapping patriarchies. As previously mentioned, Spanish-speaking women migrated north from Mexico decades, even centuries before their Euro-American counterparts ventured west. Most arrived as the wives or daughters of soldiers, farmers, and artisans. Over the course of three centuries, they raised families on the frontier and worked alongside their fathers or husbands, herding cattle and tending crops.[4]

Women's networks based on ties of blood and fictive kinship proved central to the settlement of the Spanish/Mexican frontier. At times women settlers acted as midwives to mission Indians and baptized sickly or stillborn babies. As godmothers for these infants, they established the bonds of *commadrazgo* between Native American and Spanish/Mexican women.[5] However, exploitation took place *among* women. For those in domestic service, racial and class hierarchies undermined any pretense of sisterhood. While the godparent relationship could foster ties between colonists and Native Americans, elites used baptism as a venue of social control. Indentured servitude was prevalent on the colonial frontier and persisted well into the nineteenth century.[6]

Spanish/Mexican settlement has been shrouded by myth. Walt Disney's *Zorro*, for example, epitomized the notion of romantic California controlled by fun-loving, swashbuckling rancheros. Since only 3 percent of California's Spanish/Mexican population could be considered rancheros in 1850, most women did not preside over large estates, but helped manage small family farms.[7] Married women on the Spanish/Mexican frontier had certain legal advantages not afforded their Euro-American peers. Under English common law, women, when they married, became *feme covert* (or dead in the eyes of the legal system) and thus could not own property separate from their husbands. Conversely, Spanish/Mexican women retained control of their land after marriage and held one-half interest in the community property they shared with their spouses.[8]

Life for Mexican settlers changed dramatically in 1848 with the conclusion of the U.S.–Mexican War, the discovery of gold in California, and the Treaty of Guadalupe Hidalgo. Mexicans on the U.S. side of the border became second-class citizens, divested of their property and political power. Their world turned upside down. Segregated from the Euro-American population, Mexican Americans in the barrios of the Southwest sustained their sense of identity and cherished their traditions. With little opportunity for advancement, Mexicans were concentrated in lower-echelon industrial, service, and agricultural jobs.[9] A few elite families, especially in New Mexico, retained their land and social standing. This period of conquest, physical and ideological, did not occur in a dispassionate environment. Stereotypes affected rich and poor alike, with Mexicans commonly described as lazy, sneaky, and greasy. In Euro-American journals, novels, and travelogues, Spanish-speaking women were frequently depicted as flashy, morally deficient sirens.[10]

Providing insight into community life, nineteenth-century Spanish language newspapers reveal ample information on social mores. Newspaper editors upheld the double standard. Women were to be cloistered and protected to the extent that some residents of New Mexico and Arizona protested the establishment of coeducational public schools.[11]

Despite prevailing conventions, most Mexican women, because of economic circumstances, sought employment for wages.

Whether in cities or on farms, family members pooled their earnings to put food on the table. Women worked at home taking in laundry, boarders, and sewing while others worked in the fields, in restaurants and hotels, and in canneries and laundries.[12] As sisters, cousins, and comadres, women relied on one another for mutual support. In the words of New Mexico native Fabiola Cabeza de Baca, "The women . . . had to be resourceful in every way. They were their own doctors, dressmakers, tailors, and advisers."[13] Wage work and mutual assistance were survival strategies that persisted well into the twentieth century across region and generation.

Between 1910 and 1930, over one million Mexicanos (one-eighth to one-tenth of Mexico's population) migrated "al otro lado." Arriving in the United States, often with their dreams and little else, these immigrants settled into existing communities and created new ones in the Southwest and Midwest. In 1900, from 375,000 to perhaps as many as 500,000 Mexicans lived in the Southwest. Within a short space of twenty years, Mexican Americans were outnumbered at least two to one and their colonias became immigrant enclaves. In some areas, this transformation appeared even more dramatic. Los Angeles, for example, had a Mexican population ranging from 3,000 to 5,000 in 1900. By 1930, approximately 150,000 people of Mexican birth or heritage resided in the city's expanding barrios.[14] As David Gutiérrez has so persuasively argued, immigration from Mexico in the twentieth century has had profound consequences for Mexican Americans in terms of "daily decisions about who they are—politically, socially, and culturally—in comparison to more recent immigrants from Mexico." Indeed, a unique layering of generations has occurred in which ethnic/racial identities take many forms—from the Hispanos of New Mexico and Colorado whose roots go back to the eighteenth century to the recently arrived who live as best they can in the canyons of northern San Diego County.[15]

Such a heterogeneous Mexican community is not new. Throughout the twentieth century, a layering of generations can be detected in schools, churches, community organizations, work sites, and neighborhoods. Writing about San Bernardino in the 1940s, Ruth Tuck offered the following illustration:

> There is a street . . . on which three families live side by side. The head of one family is a naturalized citizen, who arrived here eighteen years ago; the head of the second is an alien who came . . . in 1905; the head of the third is the descendant of people who

came . . . in 1843. All of them, with their families, live in poor housing; earn approximately $150 a month as unskilled laborers; send their children to "Mexican" schools; and encounter the same sort of discriminatory practices.[16]

Inheriting a legacy of colonialism wrought by Manifest Destiny, Mexicans, regardless of nativity, found themselves segmented into low-paying, low-status jobs with few opportunities for advancement. Living in segregated barrios, they formed neighborhood associations and church groups, and created a community life predicated on modes of production, economic and cultural.

This chapter surveys women's border journeys first in terms of migration and settlement followed by patterns of daily life. The ways in which women as farm worker mothers, railroad wives, and miners' daughters negotiated a variety of constraints (economic, racial, and patriarchal) are at the heart of the narrative. Mexicanas claimed a space for themselves and their families building community through mutual assistance while struggling for some semblance of financial stability, especially in the midst of rising nativist sentiments that would crest in the deportations and repatriations of the early 1930s. Whether living in a labor camp, a boxcar settlement, mining town, or urban barrio, Mexican women nurtured families, worked for wages, built fictive kin networks, and participated in formal and informal community associations. Through chain and circular migrations of families, community and kin networks intertwined. In Riverside, California, for example, the Eastside barrio by the 1960s had so many members of a single extended family that Ray Buriel recalled how he and his buddies had to venture into the rival barrio Casa Blanca to get dates.[17]

Chain and circular migrations, of course, begin with the act of crossing the political border separating Mexico and the United States.[18] In writing the history of Mexican immigration, scholars generally work within a "push/pull" model.[19] What material conditions facilitated migration and what expectations did people carry with them as they journeyed north? Between 1875 and 1910, the Mexican birthrate soared, resulting in a 50 percent increase in population. Food prices also spiraled. While dictator Porfirio Diaz has been credited with the modernization of Mexico, his economic policies decimated the lives of Mexican rural villagers as they were displaced from their *ejidos* (communal land holdings) by commercial (often corporate American) agricultural interests. Perhaps as many as five million people lost access to their ancestral lands. In the words of historian Devra Weber, "The independent Mexican peasantry disappeared, and by 1910 over nine and a half million people, 96 percent of Mexican families, were landless." By 1900, American-built and financed railroads offered mass transportation in Mexico. Since the major rail lines ran north and south (to make connections with lines on the U.S. side), hopping a train to the border was a realistic and accessible option.[20]

Beginning with the Madero uprising of 1910, the Mexican Revolution also spurred migration to the United States. Claiming the lives of an estimated one to two million people, the ten-year bloody civil war wreaked economic, political, and social chaos. Starvation was not unknown and danger a constant companion. Marauders and soldiers raped and kidnapped young women. Elsie González recalled how her grandmother had protected her sister from soldiers by throwing a wicker hamper over her and sitting on top of it until the men had left. The *soldaderas,* whether as wives, sweethearts, or paid service workers or as women who fought in their own right in their own units, shouldered multiple responsibilities in the course of a single day.[21] Although only eight years old when Diaz was routed from power in 1911, Lucia R. had clear memories of the *soldaderas:*

> They used to carry the whole house on their backs. In addition, they carried the small children and a rifle in case they had to tangle with the enemy, too. In a bucket they carried what was necessary to cook. Toward the end of the day, they would stop and set up camp and start dinner. *Pobrecitas,* they suffered a lot.[22]

Although hostilities, for the most part, would cease in 1920, the economic aftershocks reverberated throughout the following decade. In addition, the Cristero Revolt prompted further emigration from 1925 to 1929. Several scholars have referred to the United States as a "safety value" for Mexicanos seeking to escape the ravages of war. This metaphor is a good one, not only for *campesinos* and artisans, but for government officials, professionals, and the wealthy. Taking no chances, Señor Araiza, the mayor of Guadalupe I. Calvo, Chihuahua, wisely sent his wife and children to El Paso. He would never see his cherished family again as assassins would take his life during the course of the revolution.[23]

Immigrants looked to the United States as a source of hope and employment. They soon discovered that material conditions did not match their expectations. The early quantitative studies of Al-

bert Camarillo, Ricardo Romo, and Mario Barrera sharply illumi-
nated the economic and social stratification of Mexicans in the
Southwest during the early decades of the twentieth century.[24] As
examples, in 1930, the three most common occupations for Mexican men were in agriculture (45 percent), manufacturing (24 percent), and transportation (13 percent). Only 1 percent held professional positions. Women wage earners could frequently be found in
the service sector (38 percent), in blue-collar employment (25 percent), and in agriculture (21 percent). Only 3 percent were considered professionals and 10 percent held clerical or sales positions.[25]
The following discussion sketches out in the broadest strokes the
occupational niches of Mexican immigrants and their families in
the United States.

With the advent of reclamation and irrigation projects and
World War I, commercial agriculture in the Southwest boomed at
the same time that restrictive mandates against Asian immigration
contributed to "a relatively diminishing supply of workers." Growers avidly recruited Mexicanos, promising wages that seemed extraordinary to *campesinos*. Lawrence Cardoso indicated that in
Mexico, field workers could earn twelve cents per day while in the
U.S. Southwest daily wages for similar work ranged from $1.00 to
$3.50. By 1930, according to a U.S. Chamber of Commerce report,
Mexican agricultural workers earned from $2.75 to $6.00 per day.[26]
The Utah-Idaho Sugar Company contract dated March 14, 1918,
signed by Severiano Rodríguez stipulated that workers would be
paid $7.00 dollars per acre for blocking and thinning; $2.50 and
$1.50 per acre for the first and second hoeing, and $10.00 per acre
for pulling, topping, and loading sugar beets. The honoring of such
wages could be another matter altogether. In 1919, a representative of the Mexican ambassador to the United States would call on
the Commissioner General of Immigration to investigate the physical conditions of compatriots employed by the Utah-Idaho Sugar
Company based on materials the Mexican embassy had received
from Señor Rodríguez in which he explained that 500 families
"have been left in a very precarious situation."[27]

Mexicans provided the sinew and muscle on ranches and farms
throughout the West. Historian Camille Guerin-Gonzales indicates that "by 1920, Mexicans formed the largest single ethnic
group among farm workers in California, and during the 1920s,
they became the mainstay of California large-scale, specialty group
agriculture." Pioneering economist Paul Taylor found in Nueces
County, Texas, that Mexicans formed 97 percent of the farm labor

force. In Arizona, 80 percent of the year-round or "resident" farm workers were Mexican.[28] Migrating into the Pacific Northwest and Rocky Mountain states, Spanish-speaking workers could also be found in such disparate places as Nyssa, Oregon, Blackfoot, Idaho, and Green River, Wyoming. Forming over 65 percent of the sugar beet harvesters, Mexican communities also emerged in Michigan and Minnesota. The U.S. Chamber of Commerce related that by 1930 Mexicans picked "more than eighty percent of the perishable commodities produced in the Southwest."[29]

The railroads also provided employment. According to Jeff Garcílazo, Mexicanos composed from "about fifty to seventy percent of the track crews on the major western lines." Labor contractors for both agribusiness and the railroads traveled to the interior of Mexico to recruit workers holding out such inducements as high wages, free transportation, and housing. More frequently, such agents competed with one another in the border city of El Paso.[30] The border journey of the Vásquez family serves as an example. Recruited by the Rock Island Railroad in Sinalao, Guanajuato, in 1907, Felix Vásquez and his wife Frederica made their way north. Their first two children were born in Mexico and then a daughter, Euesbia, in El Paso. Laboring on the track, Vásquez with his family migrated from boxcar colonia to boxcar colonia into Arizona, New Mexico, Iowa, Kansas (where they celebrated the birth of another daughter), and then settled in Silvis, Illinois, outside Chicago, the birthplace of four younger children. The boxcar communities could move at a moment's notice or become permanent settlements. Midwest rail lines also relied on Mexican labor since over 40 percent of their workers in the Chicago–Gary region were Mexican. In 1927, wages in the rail yards of Detroit averaged $4.00 per day and Mexican rail hands could be found as far east as Pittsburgh.[31]

Mining and industrial jobs were other "pull" occupations. By 1910, Arizona had become "the nation's number one producer of copper" and the Rockefellers' Colorado Fuel and Iron Company had irrevocably altered the southern Colorado landscape with coal mines. In both states, a layering of generations occurred similar to urban areas of the Southwest with Mexicano migrants living and working alongside native-born Mexican Americans. By the mid-1920s, daily wages averaged from $2.75 to $4.95 for Mexican miners in Arizona.[32] Heavy industry in the Midwest also recruited Mexican labor, with Bethelem Steel in Pennsylvania the most notable example; in the Southwest, construction firms depended on Mexicanos. In *Mexican Immigration to the United States* (1930), anthropologist Manuel Gamio indicated that money orders to Mexico originated from such unlikely places as Nebraska and New York. The grandfather of Chicana artist Yolanda López, for example, made his living as a tailor in New York City. As Francisco Balderrama and Raymond Rodríguez astutely observed, "By the 1920s Mexicans could be found harvesting sugar beets in Minnesota, laying track in Kansas, packing meat in Chicago, mining coal in Oklahoma, assembling cars in Detroit, canning fish in Alaska, and sharecropping in Louisiana."[33]

Migration within the United States was common and the Vásquez family journey to Silvis exemplifies the stepping-stone route to the Midwest. However, most new arrivals lingered closer to the border. Coming from every Mexican state with a substantial proportional from the central plateau regions of Michoacan, Jalisco, and Guanajuato, 80 percent of this population, by 1930, lived in the states of Texas, California, New Mexico, Arizona, and Colorado.[34]

The experiences of women who journeyed north alone or only in the company of their children have received scant scholarly attention. In separate studies, however, Devra Weber and I have found numerous examples of women, like Pasquala Esparza, who arrived *al otro lado* on their own. Manuel Gamio also documents their experiences, here and there, in his field notes housed at the Bancroft Library as well as in excerpts published in his *The Life Story of the Mexican Immigrant*. The records of the Immigration and Naturalization Service, especially the transcripts of the Boards of Special Inquiry, lend insight into the lives of those who came as *solas* or as single mothers.[35]

Gender marked one's reception at the Stanton Street Bridge linking Ciudad Juárez and El Paso, especially if one ventured alone. Men would hear the competing pitches of labor contractors promising high wages and assorted benefits. Conversely, immigration inspectors routinely stopped those considered "likely to become a public charge"—in other words *solas* and single mothers. Agents scrutinized passport applications and conducted special hearings to determine women's eligibility for entrance into the United States.[36]

Arriving in Ciudad Juárez with a nine-year-old daughter and a four-week-old infant, Pasquala Esparza discovered she did not have the necessary funds to obtain the proper passports in El Paso so she stayed in Ciudad Juárez, finding a job as a housekeeper and a room in a boardinghouse. The landlady promised to look in on her

daughters while Pasquala worked; however, it was nine-year-old Jesusita who shouldered the responsibility for herself and her sister.
Jesusita remembered that as part of her daily routine she would
carry Raquel a long distance to an affluent home where their
mother worked. After preparing the noon meal for her employer,
Pasquala would anxiously wait by the kitchen door. When her children arrived, she quickly and quietly ushered them into the
kitchen. While nursing Raquel, she fed Jesusita a burrito of left-
overs. Then Jesusita would take her baby sister into her arms and
trek back to the boardinghouse to await their mother's return in
the evening. One can only imagine her fears as she negotiated the
streets of a strange city, a hungry child carrying a hungry baby. After six months, Pasquala had made enough money to complete the
journey to California.[37]

Immigration agents, however, still remained suspicious of a woman unaccompanied by a man. On their next attempt to cross,
even with cash in hand, Pasquala and her family were denied a regular passport. Desperate, but not helpless, she secured a local passport generally reserved for Juárez residents who worked in El Paso
and in that way she and her children crossed the border.[38]

Another strategy employed by women involved direct confrontation with immigration officers. Journalist John Reed recorded an incident in which a woman was queried about the contents
of her *rebozo*. "She slowly opened the front of her dress and an-
swered placidly: 'I don't know, señor. It may be a girl and it may be
a boy.'"[39] During a Board of Special Inquiry in Nogales, Arizona,
twenty-four-year-old Trinidad Orellana refused to be intimidated as
she and her fourteen-year-old sister Beatríz attempted to join their
mother and two sisters who worked as actors at the Star Theatre in
El Paso. Perhaps aware of the suspicion with which actors were
held, Trinidad adopted a defiant stance. A portion of her testimony
follows:

> ORELLANA: No, my brother is not an actor.
> HEARING OFFICER: What is he?
> ORELLANA: He is a mechanic.
> HEARING OFFICER: What kind of mechanic?
> ORELLANA: You ask him.[40]

Shortly after this exchange, an exasperated immigration agent declared, "Do you want to answer these questions . . . or do you want
to stop right now?" Appearing as a witness, her brother Alfonso took

a deferential position, emphasizing the strong transnational family bond, his fitness as a breadwinner, and his desire for U.S. citizenship. In granting the Orellana sisters admittance, the transcript reveals an odd rationale for Trinidad's testimony.

> It was thought at first from the manner of answering that there was something wrong . . . but the Board finally decided that she was just ignorant or frightened. There is nothing in her appearance to indicate that she is connected to the theatrical profession or anything other than a plain seamstress as she claims to be.[41]

Whether Trinidad Orellana's "performance" at the hearing had been carefully scripted or not, it seems interesting that she and her brother articulated reverse gender expectations—she assertive, he accommodating. As significant, immigration agents attributed her unsettling testimony to being scared or backward rather than as a direct challenge to their authority. Perhaps being caught off guard worked to the sisters' advantage, for three weeks later the El Paso office would chastise the Nogales agents for making such a hasty decision with respect to the Orellanas.[42]

The Immigration Act of 1917, which included provisions for a literacy test and a head tax, made circular migration more difficult. Historian George Sánchez contends that these measures along with harassment by border agents contributed to a pattern of more permanent settlement. Especially after its passage, immigrants arriving in El Paso (the Ellis Island for Mexicanos) encountered a daunting and demeaning reception. According to Balderrama and Rodríguez, "All immigrants, men, women, and children, were herded into crowded, examination pens. As many as five hundred to six hundred persons were detained there for endless hours without benefit of drinking fountains or toilet facilities." Immigrants were also required to remove their clothing to hand over to officials for disinfecting. They received medical examinations and were then herded through a public bath. Associating immigrants with outbreaks of influenza, border agents perceived themselves as acting in the public interest, but for the the individuals undergoing such treatment the humiliation remained a searing memory: "They disinfected us as if we were some kind of animals." Sánchez points out that this process was reserved only for poor migrants. Professional and elite exiles (and those who dressed to pass above their class) could forgo the literacy test, medical examination, and public bath.[43]

Like those who arrived from Europe and Asia, Mexican immigrants dreamed of "a better life." Some were propelled by fantastic images of prosperity. Or as a verse from a popular corrido proclaimed, "For they told me that here the dollars were scattered about in heaps; That there were girls and theaters/And that here everything was good fun."[44] The manufactured fantasies of Hollywood also appealed to adventurous young women like Elisa Silva. Divorcing an abusive husband, Silva, her mother, and two sisters left Maztalán for Los Angeles with the hope of "working as extras in the movies." However, once they arrived, they found work in different occupations. One sister worked as a seamstress, another attended business college, and Elisa earned $20 to $30 a week as a "dime a dance" partner in a local Mexican dance hall. Other women, like Pasquala Esparza, were not motivated by promises of fame and fortune; survival was their goal and, for many, the agricultural communities of Texas, California, and the Far West would be their new homes.[45]

After the grueling journey, Pasquala and her children resided with her sister's family in El Monte, California. Living under one roof with her tios and her cousins, Jesusita and her mother worked in the berry fields from February through June; then journeyed with the relatives to the San Joaquin Valley where they would first pick grapes, then cotton. By November, the extended family would return to El Monte. "We didn't work November . . . December . . . January . . . But we used to buy our sack of beans . . . and we'd get our flour and we'd get our coffee and we'd get our rice so that we could live on those three months we didn't work."[46]

It is a truism that family networks are central to American immigration history, but as I listened to Jesusita Torres, I wondered how observers, like 1930s' sociologist Ruth Allen, could have missed the complexities of extended family life when they interviewed Mexican farm workers. Indeed, Allen seemed to equate the fact that since growers paid the wages of all the family members in a lump sum to the head of household, such arrangements sat well in the minds of Mexican women, whom she believed clung to "traditions of feminine subservience." With thinly veiled contempt, Allen wrote:

> The Mexican woman has been taught as her guide to conduct the vow of the Moabitess, 'Where thou goest, I will go.' Up and down the road she follows the men of her family. . . . The modern Woman Movement and demands for economic independence

have left her untouched. Uncomplainingly, she labors in the field for months at a time and receives as a reward from the head of the family some gew-gaw from the five and ten cent store, or, at best, a new dress. The supremacy of the male is seldom disputed.[47]

The ethnocentric perceptions of this Texas professor signifies one end of the spectrum. On the opposite end reside rosy notions of happy extended families. While family and fictive kin may have eased the migrant journey and provided physical and emotional succor, human relationships are rarely perfect. Indeed, I too may be guilty of casting a fairly uncritical eye on extended family networks in *Cannery Women, Cannery Lives*.[48] Bear in mind that the dynamics of power permeate the realm of decision-making whether one is situated at work or at home. We must move beyond a celebration of la familia to address questions of power and patriarchy, the gender politics of work and family.

Gender politics, however, is also enmeshed in economic and social stratification. Women like Jesusita Torres and her mother Pasquala lived and worked in extended family relationships often by necessity rather than choice. It was not until Pasquala secured employment at a walnut factory that she could save a portion of her wages and move her daughters and grandson out of her sister and brother-in-law's home. Although Señora Torres remembered that "when you live together, you think that they love you and you love them," she also revealed that her uncle's drinking took a toll on the family. "We couldn't sleep because they had to do their singing and their cussing . . . and we had a little corner in the kitchen where we slept."[49] Julia Luna Mount remembered her family going walnut picking "with a friend of a friend." In her words, "We slept on the floor in the living room. We suffered humiliations because we really had no place to go . . . and they made us feel *very* unwelcome."[50]

Individual memories illuminate community histories. The following narrative reveals Mexican women's stories across region and occupation, examining their lives in agricultural colonias, boxcar barrios, and mining towns and focusing, in part, on the cultural construction of class. Just as women's work and family roles were intertwined, so too were the racial, economic, and patriarchal constraints they faced. Their legacies of resistance reveal their resiliency, determination, and strength.

A lifelong farm and nursery worker, Jesusita Torres stated simply:

> It's hard when you don't have an education. You go to work and
> you always have to do the hardest work. I used to think, "If I ever
> have children, I'm gonna work so hard my children will NEVER
> do this."[51]

Migrant workers, both past and present, have occupied a vulnerable, precarious sector of the working class. Indeed, as an underclass
of monopoly capitalism, frequently invisible in labor camps off the
beaten track, farm workers have, in general, labored for low wages,
under hazardous conditions, and with substandard housing and
provisions. While individual qualities such as physical stamina and
fortitude seem necessary for survival, a collective sense of family,
neighborhood, and cultural bonds created thriving colonias among
Mexican agricultural workers. In *Labor and Community*, historian
Gilbert González meticulously reconstructs citrus communities in
Orange County. Colonia residents may have depended on the growers for their livelihoods, but they developed their own local village
structures and organizations, ones imbued with what Emilio
Zamora has termed "an all-inclusive Mexicanist identity" rooted
in nationalism and "working class values of fraternalism, reciprocity, and altruism." As Devra Weber argues in *Dark Sweat, White
Gold,* "Segregation, working-class status, and the geographic mobility of Mexican men and women reinforced their identity as Mexicans . . . and reaffirmed the need to rely on each other in an Anglo-
dominated society." She continued, "While aspects of mutual aid
underlie any society, the importance of reciprocity was more powerful among immigrants."[52]

   But there is more to the story than collective identity, for the
pallor of patriarchy must also be considered in exploring the lives of
women agricultural workers. Rosalinda González contends that the
organization of farm labor reinforced patriarchal tendencies within
families. Women could labor for the *patrón* at work and the *patrón*
at home. However, like their foremothers who migrated north during the frontier era, Mexicanas created their own worlds of influence predicated on women's networks, on ties of familial and fictive
kin. *Commadrazgo* served as one of the undergirdings for general
patterns of reciprocity as women cared for one another as family
and neighbors.[53]

   As an example, Irene Castañeda recalled her mother's efforts as
a midwife in South Texas:

> Mother, from seeing the poor people die for lack of medical at-
> tention, wanted to do something to help them and she learned as
> best she could, to deliver babies. Sometimes on the floor with just

a small blanket. . . . Sometimes she would bring pillows or blankets from home—many of the women had not eaten—she would bring them rice from home and feed them by spoonfuls. The shots were a cup of hot pepper tea—to give strength for the baby to be born.[54]

The family remained the unit of production in agricultural labor. For wives and mothers, the day began before sunrise as they prepared the masa for fresh tortillas. In an interview with Gilbert González, Julia Aguirre remembered how her mother prepared tortillas on top of a steel barrel that she had improvised as a stove. As a child, Clemente Linares worked with a short-handle hoe in the beet fields of Montana. He recalled the "double day" existence of his mother who labored all day in the fields and returned to a full evening of chores. After dinner, "She would work on the washing board and tub. She had to heat the water on the stove and if there wasn't room for the water, they would heat the water outside on a fire." He continued, "She would spend half the night so she would be ready to go back to work the next morning." Drawing on a 1923 Department of Labor study, sociologists Mary Romero and Eric Margolis illuminate the double day among *campesinas* in the Colorado beet fields. "Only 14 of the 454 working mothers interviewed were relieved by other adults in the cooking and only 42 women were assisted by a child."[55]

Paid by the acre, bin, or burlap bag, workers had their earnings tied to their abilities to pick with speed and skill, careful not to bruise the berries or puncture the tomatoes. Mothers with infants were not uncommon sights. Grace Luna related how women would scale ladders with 100 pounds of cotton on their backs and "some carried their kids on top of their picking sacks."[56] While Luna picked cotton in Madera, California, María Arredondo worked in a peach orchard little more than an hour's distance near the small town of Delhi. Reflecting on her experiences as a young mother coping with the realities of migrant life, she revealed:

In 1944 we camped in Delhi under trees and orchards in tents. We made a home. We had rocks already or bricks and cooked our food and got boxes for our table. . . . Martin [her son] suffered, he remembers. Picking peaches was the hardest job—I used to cry because my neck hurt, the big peaches were heavy. I [could] only fill the bag half way because I couldn't stand the pain. . . . We lived not too far [the bosses] and that is where we used to get our water. Restrooms—they were under the trees, in the field, or by the canal.[57]

Migrant farm workers had little shelter from the extremes of heat or cold. With no labor camp in sight, Jesusita Torres dusted herself off and slept under trees. Clemente Linares recalled how the Montana winters would freeze the outdoor water pumps, but the ever-present snow, which seeped into the house from the cracks in the walls, did serve as his family's main source of water. Telling his daughter Lydia the proverbial story of walking over two miles to school in the snow, he declared, "That you didn't freeze to death was a miracle."[58] Conversely, in the poem "I Remember," Isabel Flores presents a limpid image of life in the fields on a hot summer day. A portion follows:

> I remember
> riding on my mother's
> sacka
> as she picked cotton in the middle of two
> surcos
> lonches
> tortillas y frijoles
> in an opened field
> with the dust and the wind.
>
> . . .
>
> I remember
> watching a cloud
> slowly covering the sun
> and giving thanks for the minutes
> of shade.[59]

Some children never made it to the fields. In 1938, a Michigan newspaper reported how, in the beet fields near Blissfield, company housing amounted to "hovels" with fifteen to twenty workers assigned to each shack. Babies were born "in tents or outside under trees." One infant died shortly after birth. The mother had stood in a crowded flatbed truck all the way from San Antonio, Texas, to Michigan and on her arrival went into labor prematurely.[60] A single headline from a Michigan paper says it all:

*Want, Poverty, Misery, Terror Ride Through Michigan Sugar
Beet Fields Like Four Horsemen*
Mexican Labor Brought Like Cattle to State in
Trucks; Nameless Graves Unmarked in
Fields.[61]

Rural migrant women had few choices other than picking produce. Some became cooks in labor camps and others ran makeshift
boardinghouses. In addition to picking produce, caring for her family, and serving as the local midwife, Irene Castañeda's mother took
in laundry for which she earned $5 per week. Working as a house-
keeper for local farm and merchant families offered another option, but domestic labor frequently contains the hidden psychological costs of prejudice, discrimination, and humiliation.[62] Paul
Taylor recorded the following observations from Euro-American
women in South Texas regarding their Mexican "servants."

> They are good domestic servants if you train them right. They are
> getting better and are clean if you teach them to be. . . . We feel
> toward the Mexicans like the old southerners toward the Ne-
> groes. Some of us have had servants from the same family for
> three generations.[63]

In the midst of a family tragedy, Jesusita Torres learned that she
definitely preferred migrant work over household employment. At
the age of fifteen, Jesusita eloped with a young man she met in the
fields and a year later became a mother. At seventeen, Jesusita,
pregnant with their second child, had been abandoned by her
twenty-four-year-old husband. Moving back in with her mother
and relatives, she packed carrots and spinach for a while, but then
tried working as a live-in housekeeper. Her mother would care for
her toddler son and newborn child. "I went to do housework and
they did not pay me too much and I had to stay there so I did not
like it." When Jesusita's baby died, her employer helped her provide
a proper burial. Señora Torres, however, learned that this assistance was neither an act of charity nor kindness, but an advance
she would have to pay back. In her words, "That lady helped me to
bury him because I was working for her; so after I got through paying her what I owed her then I quit."[64] How could this *patróna* be
so heartless? Writing about women of color in domestic service,
sociologist Evelyn Nakano Glenn examines both the structural
mechanism of a "dual labor system" and the playing out of racial-
ized/gendered identities and ideologies within the employer–
employee interpersonal interactions that characterize such work.
She theorizes the actions of employers in the following terms:
"Racial characterizations effectively neutralized the racial-ethnic
woman's womanhood, allowing the mistress to be "unaware" of
the domestic's relationship to her own children and household."

Nakano Glenn continues, "The exploitation of racial-ethnic women's physical, emotional, and mental work for the benefit of white
households thus could be rendered invisible in consciousness if not
in reality."[65]

Migrant women, whether they labored in the fields or someone
else's kitchen, conserved scarce familial resources within their own
households. They tended subsistence gardens and raised poultry
and other barnyard animals. At times, grandparents and children assumed responsibility for the herbs, vegetables, and chickens. Clemente Linares remembers helping his eighty-six-year-old
grandfather around the yard. "We raised tomatoes, peas, beans,
cabbage, carrots . . . in order to have a root cellar . . . to help provide us through the winter. . . . And of course, we tried to have a hog
or two to butcher, maybe a calf, and . . . we had our chickens." Such
activities lessened dependence on local merchants and the company store. As Sarah Deutsch has argued, such a mixed economy
enabled Hispanos in New Mexico and Colorado a measure of independence.[66] Yet, once they left the land, they lost that independence. Romero and Margolis explained that when these farmers
"left their dry land farms in southern Colorado or northern New
Mexico to answer the call of the growers and the sugar beet companies it was a critical step in their transformation from peasant
farmers to wage workers." They continue, "By the end of the depression the dignity of wage work had been wrested from them and
they had been reduced to underemployed wards of the state."[67]

Whether underemployed, unemployed, or even employed, putting food on the table was a full-time occupation, especially during
the Depression. In California fields, migrant farm workers of all
ethnicities (Euro-American, African American, Filipino, and Mexican) lived on the brink of starvation. John Steinbeck described a
typical diet in good times as "beans, baking powder biscuits, jam,
coffee," and, in bad, "dandelion greens and boiled potatoes." Similarly, María Arredondo recalled, "We didn't have enough food. We
had beans, very little meat mixed with potatoes and *sopa*."[68] In her
article on the San Joaquin Valley Cotton Strike of 1933, Devra Weber tellingly points to the importance of food in women's daily lives
with memories of want indelibly etched in their consciousness.
"Men remembered the strike in terms of wages and conditions;
women remembered the events in terms of food."[69]

For some, resistance to exploitation took the form of labor activism; for others, escape seemed the only option. A single case
study taken from INS records can serve to show fortitude and
courage. It concerns over 150 Mexican immigrants recruited to

pick sugar beets by the Utah-Idaho Sugar Company only to find that management failed to abide by the terms of their contracts and the recruited immigrants were left to fend for themselves without coal or food in the bleak Idaho winter. As mentioned earlier, one of the workers, Severiano Rodríguez, had appealed to the Mexican ambassador to intervene on their behalf. In addition, a local priest brought their plight to the county commissioners who authorized the distribution of 1,000 pounds of flour and one ton of coal as well as a relief allotment of $165 to be divided among sixteen of the neediest families. The county commission then sought compensation from the sugar beet firm. A subsequent Immigration Service investigation absolved the Utah-Idaho Sugar Company from any wrongdoing. Referring to company representatives as "intelligent and capable men," the investigating agent believed that incidences of suffering had been "exaggerated." In a classic example of scapegoating, he chastised Mexicanos for not bringing along the proper clothes and bedding for an Idaho winter and not saving enough of their wages to carry them through to spring. Although he realized that the workers would be charged for such supplies, he seemed bewildered that they turned down company offers of blankets and mattresses. The migrants had already accumulated substantial debt, beginning with company charges for transportation to Idaho, and, though in great want, they were determined to avoid further employer claims to their labor. During the investigation, the Utah-Idaho Sugar Beet Company also made assurances that the Mexican immigrants would henceforth receive adequate supplies of food and fuel. Although the local government had donated some provisions, Mexicans were not welcome as they were perceived as carriers of influenza and even the Immigration Service acknowledged that at least seven migrants had succumbed to the epidemic.[70]

Having little recourse and probably fewer resources, thirty-two people—men, women, and children—gathered their belongings and fled the labor camp. Like the African-American slaves who took a chance on the Underground Railroad, these Mexicano immigrants (twenty-one were members of a single extended family, the Betancourts) made a desperate break for freedom. Behaving like a modern-day planter, the Utah-Idaho Sugar Company appealed to immigration authorities for assistance in apprehending those whom the firm perceived as breaking their contracts with their feet. Labeling them deserters, a company official wrote, "I understand that some of the people are in Pocatello, Idaho, but have reason to believe some of them have gone to Elko, Nevada."[71]

Resistance to economic exploitation could also take the form of ethnic community building. In the citrus belt of southern California, Mexican immigrants established colonias or villages complete with their own organizations and institutions. Forming patriotic associations, mutual aid groups, church societies, and baseball teams, Mexican immigrants created a rich, semiautonomous life for themselves. In historian Gilbert González's words, "The village was home, neighborhood, playground, and social center." The length of the citrus season promoted the development of Orange County colonias and Riverside barrios. With employment available in the groves eight months out of the year, citrus workers had a spatial stability in contrast to transient or contract labor. During the off-season (late summer, early fall), citrus families would often make the migrant circuit north picking grapes and cotton in the San Joaquin Valley or perhaps heading southeast to the rich agricultural fields near Coachella. However, they had a home and community awaiting their return.[72] For Eusebia Vásquez de Buriel, Our Lady of Guadalupe Shrine has been at the center of her life for over sixty years. She recalled how the Mexican neighbors chipped in to build their own church in the middle of the Depression. "We worked real hard to have our church . . . the people were all poor, worst than we are now, but everything came up real nice, so we are very proud of . . . that church." Citrus communities represented a collective identity and a sense of belonging for its members or, as Gilbert González stated, within these villages, workers "constructed their vision of a good society."[73]

Conditions of migrant life were not confined to agricultural labor. Railroad workers and their families traveled from one boxcar barrio to another. While men went off to the tracks, women endeavored to make the boxcar a home and to nurture ties with their neighbors. When newcomers arrived in Belen, New Mexico, for example, women met the crew trains offering their assistance to the passengers. Frederica Vásquez recalled to her grandson Ray Buriel how "las señoras . . . went out to meet them and brought them food and brought them clothing and made them feel very welcome."[74]

According to historian Jeffrey Garcílazo, "Boxcar communities probably represented the most common form of housing for Mexican workers and their families."[75] Some were "rolling villages" in that families traveled with their particular shelters while other settlements were composed of boxcars with the wheels removed. The company provided wood-burning stoves and at times outdoor sanitation facilities. However, one Kansas man stated that the outhouse only had two seats for thirty people. Given the isolation of many of these settlements, families often had little choice but to buy their staples from the company commissary. Like the contract laborers employed by the Utah-Idaho Sugar Company, track families could become entangled in a web of debt peonage. Noting the high prices charged by the company store, the wife of a Southern Pacific rail worker, Juana Calderón, declared, "We cannot save any money . . . always in debt so we will probably always stay with the . . . Company."[76]

As in the case of farm workers, subsistence gardens and barnyard animals could ease reliance on the company store as well as provide fresh produce, dairy products, and meat. Women and children tended the chickens and goats, pulled weeds, and nurtured seedlings.[77] One can only imagine their frustration if their husbands were transferred to another section as they gathered their belongings and livestock, leaving the gardens for other families to harvest.

A single boxcar often housed more than one family, generally two, sometimes more. Families sweltered in the Arizona heat and shivered in an Illinois winter. Referring to the railroad settlement of Silvis, Illinois, a *Reader's Digest* article related: "When the Mexicans in their boxcars woke up in the wintertime, children had to break ice in the washbowls before they could clean up for school."[78] Health care, moreover, was a vital concern. People frequently relied on *curas,* those in the community with knowledge of traditional medicine. Doctors and hospitals were not readily accessible. Frederica Vásquez would lose two daughters in Silvis, one to whooping cough and another to pneumonia.[79]

Railroad wives, like migrant workers, could also find making ends meet a difficult proposition. To supplement their spouses' incomes, they took in sewing, laundry, boarders, even babies. Some women earned money or food for their families by wet-nursing neighborhood infants. As Gregoria Sosa, a railroad worker's wife from Colton, California, recounted:

> I bore three children and did washing and ironing for some of my neighbors. Sometimes I was also a wet nurse. I was very sad once when one of my "criados"—a child I breast fed was taken from my breast because his father did not want to pay me any longer. The baby died of hunger not much later. They tried to have him suck on a goat teat. I would have fed him without money, for a little food to help my little ones.[80]

Seeking some measure of economic security, railroad workers in Silvis, Illinois, "saved enough money to buy land that no else wanted at the west end of town." These men had relatively stable jobs in the repair shop of the Rock Island Railroad.[81] After years of migrating on the Rock Island rails, Felix Vásquez would also secure employment as a bolt maker at the Silvis plant and for ten years he and his family would call Silvis home. Recalling the close-knit nature of the community, his daughter Eusebia stated, "Most people were real nice, they called [each other] los compadres." She further explained that during the influenza epidemic of 1918, her father organized a food drive to assist his afflicted neighbors. "My father used to have a little wagon and every week, he used to go to every house and pick up food . . . to help the sick people."[82]

Mutual aid proved a cornerstone in the process of settlement among Mexican workers in the United States. It should be noted that not everyone participated in this sense of reciprocity, as evident in Gregoria Sosa's narrative. However, like the frontier women described by Fabiola Cabeza de Baca in *We Fed Them Cactus*,[83] Mexicanas, whether in migrant camps, boxcar barrios, or mining towns, sought to exercise some control over their lives, often relying on one another for material and emotional support.

The cultural construction of class can be discerned in the mining communities of southern Arizona and southern Colorado. Both locales had a mixed economy—mining towns next to villages with ranches and homesteads marking the landscape. In *Songs My Mother Sang to Me*, Patricia Preciado Martin presents the oral narratives of ten Arizona women, women whose memories elucidate the division of labor within families as well as the layering of generations within a regional matrix. Furthermore, Martin's narrators demonstrate how women claimed a public space through expressions of religious faith.[84]

Typical of working-class Mexican and Mexican-American households, the family served as the locus of production. Whether from a ranching or mining family, daughters were expected to perform a round of arduous chores. The labor of female kin, regardless of age, proved instrumental in ensuring the family's economic survival. Women preserved food for the winter, sold surplus commodities to neighbors, did laundry for Euro-American employers, and provided homes for lodgers. Like their pioneer foremothers, they also herded livestock, milked cows, built fences, and harvested crops. A strict division of labor according to gender became blurred. Yet this seemingly egalitarian assignment of tasks in no way subverted the traditional notion of "woman's place." Before the break of dawn, Rosalía Salazar and her sisters would rise to gather kindling, milk the cows, and afterwards walk sever~ to school, a routine that began with serving their father a coffee.[85]

With fortitude, faith, and unsung courage, single mothers relied on their domestic skills to feed their children. Julia Yslas Vélez recalled how her mother, who came from a middle-class background in Mexico, peddled her handmade garments to "poor" Mexicanos. "She did not have a formal education, but she was very smart. She had a little book. . . . She used to mark in it what people owed her. She would draw a circle for a dollar and a half circle for fifty cents."[86] Across Arizona and the Southwest, women participated in the informal economy in various ways—lodging single miners in Superior, Arizona, selling pan dulce door to door in San Bernardino, or swapping sex for food in El Paso. Some relied on their healing skills. As *curanderas* (healers) and *pateras* (midwives), Mexican women nurtured the networks essential for claiming a place in the United States.

A layering of generations and peoples characterized rural Arizona. Mexicano migrants from Sonora homesteaded alongside Mexican Americans. Marriages occurred across generational and racial lines. Boardinghouses brought people together. At Josepha's Boarding House in Superior, for instance, a young Sonoran miner successfully courted Josefa's Arizona-born daughter. The oral histories in *Songs My Mother Sang to Me* reveal a multiracial agrarian society. As an example, Rosalía Salazar was the child of a Mexican mother and a "full-blooded Opata Indian" father. She married Wilford Whelan, whose mother Ignacia was Mexicana.[87] In the center of this multiracial society was a distinctive Mexican-American agrarian culture, one that incorporated those willing to partake of it. Some "Americanos" attended fiestas, dances, and religious pageants. Assimilation was a not a one-way street. In southern Arizona, assimilation seemed to be thrown in reverse. Intermarriage did not guarantee the anglicization of the region's Spanish-speaking peoples. "Many of the offspring of Mexican-Anglo unions emphasized their Mexican rather than their Anglo heritage," observed historian Thomas Sheridan. "The reasons they did so testify to the enduring strength of Mexican society in the face of Anglo political and economic hegemony."[88] One also has to take into account the class bridge, with Mexican-Euro-American intermarriage occurring among those who owned property. The voices represented in *Songs* point to an expansive Mexican cultural horizon in southern Arizona where one's "positionality" or identity rested not in some essentialist biological mooring but through acceptance and adoption of Mexican cultural values and expectations.

Yet southern Arizona was a stratified society complete with segregated schools and clearly demarcated "American" and "Mexican" sides of mining towns. "I'll admit there was a lot of discrimination in those years," declared Carlotta Silvas Martin as she recalled growing up as a miner's daughter. In Mascot, Dolores Montoya opened a boardinghouse in the Euro-American section of the town. Decades later, her daughter Esperanza would vividly recount the fear she felt as she, her recently widowed mother, and her siblings were forced to abandon the family-run boardinghouse in the face of systematic terror and harassment. In the dark of night, someone kept turning the doorknob and separating the vines from the window. Reaching a point of desperation, the family fled with only their clothing "After we left, whoever it was did a good job of robbing us. They took everything—dishes, jewelry, furniture—anything of value, even the *santos*."[89]

Women relied on one another and on their faith. Religious practices permeated everyday routines. In preparing the masa for the tortillas, María del Carmen Trejo de Gastelum "would always add salt to the flour in the form of *la Santa Cruz* (the Holy Cross)— *para bendecir la masa* (to bless the dough)." With regard to education, the convent of the Sisters of the Company of Mary in Douglas, Arizona, served as a bulwark against the Americanizing influences of a mining town. The nuns became teachers of both catechism and custom.[90] Church jamaicas, saints' days, and Mexican patriotic holidays constituted an integral part of Arizona's Mexican-American agrarian culture. Recalling the celebration of "Las Posadas," Carlotta Silvas Martin observed:

> Las Posadas are a reenactment of the travels of Joseph and Mary who are looking for shelter before the birth of Jesus. Large groups of men, women, and children walked in procession thorough the darkened streets carrying candles. . . . We'd arrive at a designated house and sing songs asking for *posada* or lodging . . . those inside would answer that there was no room. We'd go to several houses until we arrived at a chosen house. . . . Then we'd go in and have food—chocolate and *pan de huevo* . . . and a piñata full of candy.[91]

Las Posadas reaffirmed the practice of ritualized visiting among kin and friends; it seemed as much a celebration of community networks as a religious journey. From a small home altar nestled atop a bureau dresser to a well-orchestrated town play or pageant, Mexicans in southern Arizona viewed their own interpretations of Catholicism as integral parts of their cultural life. Women also carved out a public cultural space in these community-based religious productions.

Women's daily lives appear to corroborate Richard White's observation on the cultural construction of class. "A self-conscious working class demands not just common labor, but also a common sense of identity, a common set of interests, and a common set of values." Arguing that "ethnic solidarity often seemed more important than working class solidarity," White maintained that Western mining towns "often seemed a collection of separate ethnic working-class communities whose overarching class consciousness was tentative and fragile when it existed at all."[92]

While racial/cultural boundaries could blur in Arizona's agrarian communities, in southern Colorado, ethnic boundaries appear relatively fixed, with racial/class divisions cropping up even within groups of Spanish-speaking workers. Born in Walsenberg, Colorado, in 1921, Erminia Ruiz was considered the daughter of a "mixed marriage"—her father was a Mexican immigrant, her mother a Hispana born in nearby Trinidad. She remembered that Mexican union families (those associated with the Industrial Workers of the World) tended to stick together. On Saturday night, they would gather at someone's house for music, food, dancing, and fellowship. "All the neighbors got together. You'd have dancing and they put all the chairs out . . . and the ladies would bake pies and cakes." During the Columbine Strike of 1927, Erminia had little contact with her mother's side of the family as her uncles were scabs. "We were in a way closer to our neighbors." She also remembered attending union meetings with her father, sitting on his knee and listening to all the languages spoken around her. There, she learned to sing her first song in English—"Solidarity Forever." Her personal story correlates well with Sarah Deutsch's analysis of the ways in which ethnic and regional identities in New Mexico and Colorado reconfigure class consciousness within separate communities.[93]

Whether they lived in a camp, village, or city, Mexican women carved a place for themselves and their families based on shared experiences, cultural traditions, histories, and concerns. They relied on one another as family members and as neighbors whether they lived in a tightly knit rural colonia or a rolling boxcar barrio. Yet, as we have seen, patriarchy and even class distinctions existed; families could be source of strength or a source of trial. But the range of their lives and their struggles seemed lost on the American public. Growing nativist sentiment during the 1920s and 1930s began to blame Mexican immigrants for society's ills.

A Mexican "expert" from Vanderbilt University, Dr. Roy Garis testified before a U.S. congressional committee. He reiterated the views of a Euro-American Westerner, a man who claimed that Mexican women were instinctively prone to adultery. Relaying this questionable third-party testimony, Garis recapitulated the tired, trite, and grotesque nineteenth-century gendered, racialized stereotypes for a modern audience.[94] A portion follows:

> Their minds run to nothing higher than animal functions—eat, sleep, and sexual debauchery. In every huddle of Mexican shacks one meets the same idleness . . . filthy children with faces plastered with flies, diseases, lice . . . apathetic peons and lazy squaws.[95]

These sentiments were not isolated, extremist meanderings. With a circulation of nearly three million, *The Saturday Evening Post* ran a series of articles urging the restriction of Mexican immigration. The titles tell the story: "The Mexican Invasion," "Wet and Other Mexicans," and "The Alien on Relief." One article, "The Docile Mexican," characterized Mexicano immigrants with the following adjectives: "illiterate, diseased, pauperized." Relying on mixed metaphors as well as the opinions of scientists who dabbled in eugenics, the author Kenneth Roberts refers to Mexicans as both "white elephants" and as people who bring "countless numbers of American citizens into the world with the reckless prodigality of rabbits." Roberts cautions against "the mongrelization of America," warning further that the children of Mexican and Euro-American parents will result in "another mixed race problem; and as soon as a race is mixed, it is inferior."[96] And under the heading of "The Mexican Conquest," the editor of *The Saturday Evening Post* offered his opinion in the June 22, 1929, issue:

> The very high Mexican birth rate tends to depress still further the low white birth rate. Thus a race problem of the greatest magnitude is being allowed to develop for future generations to regret and in spite of the fact that the Mexican Indian is considered a most undesirable ethnic stock for the melting pot.[97]

With the onset of the Great Depression, rhetoric exploded into action. Between 1931 to 1934, an estimated one-third of the Mexican population in the United States (over 500,000 people) were either deported or repatriated to Mexico even though the majority were native U.S. citizens. Mexicans were the only immigrants targeted for removal. Proximity to the Mexican border, the physical distinctiveness of mestizos, and easily identifiable barrios influenced immigration and social welfare officials to focus their efforts solely on the Mexican people, people whom they viewed as both foreign usurpers of American jobs and as unworthy burdens on relief rolls. From Los Angeles, California, to Gary, Indiana, Mexicans were either summarily deported by immigration agencies or persuaded to depart voluntarily by duplicitous social workers who greatly exaggerated the opportunities awaiting them south of the border.[98] In the words of George Sánchez,

> As many as seventy-five thousand Mexicans from southern California returned to Mexico by 1932. . . . The enormity of these figures, given the fact that California's Mexican population was in 1930 slightly over three hundred and sixty thousand . . . indicates that almost every Mexican family in southern California confronted in one way or another the decision of returning or staying.[99]

Francisco Balderrama and Raymond Rodríguez place the deportation and repatriation figures even higher. Drawing on statistics from both U.S. and Mexican government agencies as well as newspaper reports, they contend that one million Mexicanos were repatriated or deported during the 1920s and 1930s. Moreover, they note "that approximately 60 percent . . . were children who had been born in the United States."[100]

The methods of departure varied. A historian of Los Angeles, Douglas Monroy, recounts how *la migra* trolled the barrio in a "dog catcher's wagon." In one instance, immigration agents tore a Los Angeles woman from her home in the early morning hours, threw her in the wagon, and then left her toddler screaming on the front porch.[101] Even if such scenes were few and far between, they certainly invoked fear among Mexicanos, many of whom decided to take the county up on its offer of free train fare. Carey McWilliams described those boarding a repatriation train as "men, women, and children—with dogs, cats, and goats . . . [with] half-open suitcases, rolls of bedding, and lunch baskets."[102] Thousands more chose to leave by automobile. They piled all their possessions—mattresses, furniture, clothing—into a jalopy and headed south. This scene of auto caravans making their way into the interior of Mexico offers a curious parallel to the ensuing Dust Bowl or "Okie" migration into California.[103]

Losing one child and struggling to support the other, Jesusita Torres held on to her place in the United States. She refused to apply for relief because she and her mother wanted to escape the notice of government authorities. "My mother said it was no use for us to [go] back . . . to what? We did not have anything out there." Describing the repatriation of two friends, she further remarked, "We were sorry that they left, because both of the ladies the husbands left them [in Mexico] with their children. It was pretty hard for them." Jesusita survived the Depression by picking berries and string beans around Los Angeles and following the crops in the San Joaquin Valley. From her wages, she raised a family and bought a house, one she purchased for seventeen dollars.[104]

Petra Sánchez had no choice. By the fall of 1933, Petra and Ramón appear to have built a nice life for themselves in Buena Park. With the money from berry picking and manure hauling combined with Petra's frugal budgeting, the couple had leased a small ranch. From 1926 to 1933, their family grew from four children to ten.[105] According to Marjorie Sánchez-Walker,

> Even with a new baby arriving every fifteen months, Petra still found the time to supplement the family's needs from her industry. Chicken provided eggs and meat that she could sell when there was a surplus; her garden produced vegetables; she made cheeses which hung . . . over the dining room table; and in the summer, she picked berries for wages.[106]

Petra found she could not keep up this pace. In November 1933, she suffered a nervous breakdown and was committed to the Norwalk State Mental Hospital.[107]

By Christmas, Petra, her health seemingly restored, would be home with her family again, but home now was her childhood village of San Julián. Coming under the scrutiny of relief authorities, Ramón believed that if the family left voluntarily, they could return at a later date. However, his papers bore the stamp: "LOS ANGELES COUNTY/DEPARTMENT OF CHARITIES/COUNTY WELFARE DEPARTMENT." The family now bore the onus of "liable to become a public charge" and thus "ineligible for readmission." "Repatriation, therefore, amounted to deportation for Petra and Ramón."[108]

In 1935, hoping to return to California, the couple and their eleven children, with another on the way, traveled to Ciudad Juárez. They were turned away at the border. With money running low, Ramón "shaved his moustache, borrowed money for a second-hand suit and with his green eyes and fair-skin, simply walked across the border." He planned to earn enough money picking berries in Buena Park to secure his family's clandestine passage to the United States. The children supported the family—the boys by shining shoes, selling trinkets, and "lagging pennies"; and the girls by running errands for neighbors. Six-year-old Juan acted as a "tour guide" for U.S. army personnel on the prowl for a good time in the Red Light District and for his labor received tips from both soldiers and prostitutes.[109]

The deprivation in Ciudad Juárez was well known. The *New York Times* carried a story of how over twenty repatriates had died "from pneumonia and exposure." Without resources or shelter, "as many as 2,000 lived in a large open corral."[110] With hunger a constant companion, Petra gave birth to a daughter Catalina, but the infant would die in Juárez fifteen months later, her coffin handmade by her brother Librado. Petra held her children together under the most adverse circumstances. In 1937, the family was reunited in California; but Ramón and Petra would never regain the level of financial security they had known living on their leased Buena Park ranch.[111]

After 1934, the deportation and repatriation campaigns diminished, but the effects of the Depression, segregation, and economic segmentation remained. Even members of the middle-class Mexican-American community were not immune. During the 1930s and 1940s, the League of United Latin American Citizens (LULAC) led the fight for school desegregation in the courts. At the household level, maintaining appearances proved important. With no money for coal, Eduardo Araiza, who owned a small auto repair shop in El Paso, brought home rubber tires to burn in the fireplace. As his daughter Alma related, "You kept up appearances even though your stomach grumbled."[112]

The border journeys of Mexican women were fraught with unforeseen difficulties, but held out the promises of a better life. In the words of one Mexicana, "Here woman has come to have a place like a human being."[113] Women built communities of resiliency, drawing strength from their comadres, their families, and their faith. Confronting "America" often mean confronting the labor contractor, the boss, the landlord, or *la migra*. It could also involve negotiating the settlement house, the grammar school, and the health clinic. State and church-sponsored Americanization projects could portend cultural hegemony, individual empowerment, vocational tracking, community service, or all four simultaneously. To get at how Mexicanas and their children traversed the terrain of Americanization in negotiating institutions and ideologies, a case study seems appropriate. The Rose Gregory Houchen Settlement House in El Paso, Texas, emphasized "Christian Americanization" while furnishing social services denied Mexicans in the public sector. A historical survey of this Methodist settlement reveals much about how women, especially as mothers and daughters, claimed portions of Americanization within their own cultural frames.

2

Confronting "America"

❖❖❖

As a child Elsa Chávez confronted a "moral" dilemma. She wanted desperately to enjoy the playground equipment close to her home in El Paso's Segundo Barrio. The tempting slide, swings, and jungle gym seemed to call her name. However, her mother would not let her near the best playground (and for many years the only playground) in the barrio. Even a local priest warned Elsa and her friends that playing there was a sin—the playground was located within the yard of the Rose Gregory Houchen Settlement, a Methodist community center.[1]

While one group of Americans responded to Mexican immigration by calling for restriction and deportation, other groups mounted campaigns to "Americanize" the immigrants. From Los Angeles, California, to Gary, Indiana, state and religious-sponsored Americanization programs swung into action. Imbued with the ideology of "the melting pot," teachers, social workers, and religious missionaries envisioned themselves as harbingers of salvation and civilization.[2] Targeting women and especially children, the vanguard of Americanization placed their trust "in the rising generation." As Pearl Ellis of the Covina City schools explained in her 1929 publication, *Americanization Through Homemaking,* "Since the girls are potential mothers and homemakers, they will control, in a large measure, the destinies of their future families." She continued, "It is she who sounds the clarion call in the campaign for better homes."[3]

33

A growing body of literature on Americanization in Mexican communities by such scholars as George Sánchez, Sarah Deutsch, Gilbert González, and myself suggest that church and secular programs shared common course offerings and curricular goals. Perhaps taking their cue from the regimen developed inside Progressive Era settlement houses, Americanization projects emphasized classes in hygiene, civics, cooking, language, and vocational education (e.g., sewing and carpentry). Whether seated at a desk in a public school or on a sofa at a Protestant or Catholic neighborhood house, Mexican women received similar messages of emulation and assimilation. While emphasizing that the curriculum should meet "the needs of these people," one manual proclaimed with deepest sincerity that a goal of Americanization was to enkindle "a greater respect . . . for our civilization."[4]

Examples of Americanization efforts spanned the Southwest and Midwest from secular settlements in Watts, Pasadena, and Riverside to Hull House in Chicago. In addition, Catholic neighborhood centers, such as Friendly House in Phoenix, combined Americanization programs with religious and social services. Protestant missionaries, furthermore, operated an array of settlements, health clinics, and schools. During the first half of the twentieth century, the Methodist Church sponsored one hospital, four boarding schools, and sixteen settlements/community centers, all serving a predominately Mexican clientele. Two of these facilities were located in California, two in Kansas, one in New Mexico, and sixteen in Texas.[5] Though there are many institutions to compare, an overview, by its very nature, would tend to privilege missionary labors and thus, once again, place Mexican women within the shadows of history. By taking a closer look at one particular project—the Rose Gregory Houchen Settlement—one can discern the attitudes and experiences of Mexican women themselves. This chapter explores the ways in which Mexican mothers and their children interacted with the El Paso settlement, from utilizing selected services to claiming "American" identities, from taking their babies to the clinic for immunizations to becoming missionaries themselves.

Using institutional records raises a series of important methodological questions. How can missionary reports, pamphlets, newsletters, and related documents illuminate the experiences and attitudes of women of color? How do we sift through the bias, the self-congratulation, and the hyperbole to gain insight into women's lives? What can these materials tell us of women's agencies within and against larger social structures? I am intrigued (actually obsessed is a better word) with questions involving decision-making, specifically with regard to acculturation. What have Mexican women chosen to accept or reject? How have the economic, social, and political environments influenced the acceptance or rejection of cultural messages that emanate from the Mexican community, from U.S. popular culture, from Americanization programs, and from a dynamic coalescence of differing and at times oppositional cultural forms? What were women's real choices and, to borrow from Jürgen Habermas, how did they move "within the horizon of their lifeworld"?[6] Obviously, no set of institutional records can provide substantive answers, but by exploring these documents through the framework of these larger questions, we place Mexican women at the center of our study, not as victims of poverty and superstition as so often depicted by missionaries, but as women who made choices for themselves and for their families.

As the Ellis Island for Mexican immigrants, El Paso seemed a logical spot for a settlement house. In 1900, El Paso's Mexican community numbered only 8,748 residents, but by 1930 this population had swelled to 68,476. Over the course of the twentieth century, Mexicans composed over one-half the total population of this bustling border city.[7] Perceived as cheap labor by Euro-American businessmen, they provided the human resources necessary for the city's industrial and commercial growth. Education and economic advancement proved illusory as segregation in housing, employment, and schools served as constant reminders of their second-class citizenship. To cite an example of stratification, from 1930 to 1960, only 1.8 percent of El Paso's Mexican workforce held high white-collar occupations.[8]

Segundo Barrio or South El Paso has served as the center of Mexican community life. Today, as in the past, wooden tenements and crumbling adobe structures house thousands of Mexicanos and Mexican Americans alike. For several decades, the only consistent source of social services in Segundo Barrio was the Rose Gregory Houchen Settlement House and its adjacent health clinic and hospital.

Founded in 1912 on the corner of Tays and Fifth in the heart of the barrio, this Methodist settlement had two initial goals: (1) provide a Christian roominghouse for single Mexicana wage earners and (2) open a kindergarten for area children. By 1918, Houchen offered a full schedule of Americanization programs—citizenship, cooking, carpentry, English instruction, Bible study, and Boy

Scouts. The first Houchen staff included three Methodist missionaries and one "student helper," Ofilia [sic] Chávez.[9] Living in the
barrio made these women sensitive to the need for low-cost, accessible health care. Infant mortality in Segundo Barrio was alarm-
ingly high. Historian Mario García related the following example:
"Of 121 deaths during July [1914], 52 were children under 5 years
of age."[10]

Houchen began to offer medical assistance, certainly rudimen-
tary at first. In 1920, a registered nurse and Methodist missionary
Effie Stoltz operated a first aid station in the bathroom of the settlement. More important, she soon persuaded a local physician to
visit the residence on a regular basis and he, in turn, enlisted the
services of his colleagues. Within seven months of Stoltz's arrival, a
small adobe flat was converted into Freeman Clinic. Run by volunteers, this clinic provided prenatal exams, well-baby care, and pedi-
atric services and, in 1930, it opened a six-bed maternity ward.
Seven years later, it would be demolished to make way for the construction of a more modern clinic and a new twenty-two-bed maternity facility—the Newark Methodist Maternity Hospital. Health
care at Newark was a bargain. Prenatal classes, pregnancy exams,
and infant immunizations were free. Patients paid for medicines at
cost and, during the 1940s, $30 covered the hospital bill. Staff
members would boast that for less than $50, payable in install-
ments, neighborhood women could give birth at "one of the best
equipped maternity hospitals in the city."[11]

Houchen Settlement also thrived. From 1920 to 1960, it coordinated an array of Americanization activities. These included age
and gender graded Bible studies, music lessons, Campfire activities, scouting, working girls' clubs, hygiene, cooking, and citizenship. Staff members also opened a day nursery to complement the
kindergarten program. In terms of numbers, how successful was
Houchen? The available records give little indication of the extent
of the settlement's client base. Based on fragmentary evidence for
the period 1930 to 1950, perhaps as many as 15,000 to 20,000
people per year or approximately one-fourth to one-third of El
Paso's Mexican population utilized its medical and/or educational
services. Indeed, one Methodist from the 1930s pamphlet boasted
that the settlement "reaches nearly 15,000 people."[12]

As a functioning Progressive Era settlement, Houchen had
amazing longevity from 1912 to 1962. Several Methodist missionaries came to Segundo Barrio as young women and stayed until
their retirement. Arriving in 1930, Millie Rickford would live at the

settlement for thirty-one years. Two years after her departure, the Rose Gregory Houchen Settlement House (named after a Michigan schoolteacher) would receive a new name, Houchen Community Center. As a community center, it would become more of a secular agency staffed by social workers and at times Chicano activists.[13] In 1991 the buildings that cover a city block in South El Paso still furnish day care and recreational activities. Along with Bible study, there are classes in ballet folklorico, karate, English, and aerobics. Citing climbing insurance costs (among other reasons), the Methodist Church closed the hospital and clinic in December 1986 over the protests of local supporters and community members.[14]

From 1912 until the 1950s, Houchen workers placed Americanization and proselytization at the center of their efforts. Embracing the imagery and ideology of the melting pot, Methodist missionary Dorothy Little explained:

> Houchen settlement stands as a sentinel of friendship . . . between the people of America and the people of Mexico. We assimilate the best of their culture, their art, their ideals and they in turn gladly accept the best America has to offer as they . . . become one with us. For right here within our four walls is begun much of the "Melting" process of our "Melting Pot."[15]

The first goal of the missionaries was to convert Mexican women to Methodism since they perceived themselves as harbingers of salvation. As expressed in a Houchen report, "Our Church is called El Buen Pastor . . . and that is what our church really is to the people—it is a Good Shepherd guiding our folks out of darkness and Catholocism [sic] into the good Christian life." Along similar lines, one Methodist pamphlet printed during the 1930s equated Catholicism (as practiced by Mexicans) with paganism and superstition. Settlement's programs were couched in terms of "Christian Americanization" and these programs began early.[16]

Like the Franciscan missionaries who trod the same ground three centuries before, Houchen settlement workers sought to win the hearts and minds of children. While preschool and kindergarten students spoke Spanish and sang Mexican songs, they also learned English, U.S. history, biblical verses—even etiquette a la Emily Post.[17] The settlement also offered various after-school activities for older children. These included "Little Homemakers," scouting, teen clubs, piano lessons, dance, bible classes, and story hour. For many years the most elaborate playground in South El

Paso could be found within the outer courtyard of the settlement. Elsa Chávez eventually got her playground wish. She and her mother reached an agreement: Elsa could play there on the condition that she not accept any "cookies or koolaid," the refreshments provided by Houchen staff. Other people remembered making similar bargains—they could play on the swings and slide, but they could not go indoors.[18] How big of a step was it to venture from the playground to story hour?

Settlement proselytizing did not escape the notice of barrio priests. Clearly troubled by Houchen, a few predicted dire consequences for those who participated in any Protestant-tinged activities. As mentioned earlier, one priest went so far as to tell neighborhood children that it was a sin even to play on the playground equipment. Others, however, took a more realistic stance and did not chastise their parishioners for utilizing Methodist child care and medical services. Perhaps as a response to both the Great Depression and suspected Protestant inroads, several area Catholic churches began distributing food baskets and establishing soup kitchens.[19]

Children were not the only ones targeted by Houchen. Women, particularly expectant mothers, received special attention. Like the proponents of Americanization programs in California, settlement workers believed that women held a special guardianship over their families' welfare. As head nurse Millie Rickford explained, "If we can teach her [the mother-to-be] the modern methods of cooking and preparing foods and simple hygiene habits for herself and her family, we have gained a stride."[20]

Houchen's "Christian Americanization" programs were not unique. During the teens and twenties, religious and state-organized Americanization projects aimed at the Mexican population proliferated throughout the Southwest. Although these efforts varied in scale from settlement houses to night classes, curriculum generally revolved around cooking, hygiene, English, and civics. Music seemed a universal tool of instruction. One Arizona schoolteacher excitedly informed readers of *The Arizona Teacher and Home Journal* that her district for the "cause of Americanization" had purchased a Victorola and several records that included two Spanish melodies, the "'Star Spangled Banner,' 'The Red, White, and Blue,' 'Silent Night,' . . . [and] 'Old Kentucky Home.'"[21] Houchen, of course, offered a variety of musical activities beginning with the kindergarten rhythm band of 1927. During the 1940s and 1950s, missionaries provided flute, guitar, ballet, and tap lessons. For fifty cents a week, a youngster could take dance or music classes and perform in settlement recitals.[22] Clothing youngsters in European peasant styles was common. For instance, Alice Ruiz, Priscilla Molina, Edna Parra, Mira Gómez, and Aida Rivera represented Houchen in a local Girl Scout festival held at the Shrine temple in which they modeled costumes from Sweden, England, France, Scotland, and Lithuania.[23] Some immigrant traditions were valorized more than others. Celebrating Mexican heritage did not figure into the Euro-American orientation pushed by Houchen residents.

In contrast, a teacher affiliated with an Americanization program in Watts sought to infuse a multicultural perspective as she directed a pageant with a U.S. women's history theme. Clara Smith described the event as follows:

> Women, famous in the United States history as the Pilgrim, Betsy Ross, Civil War, and covered wagon women, Indian and Negro women, followed by the foreign women who came to live among us were portrayed. The class had made costumes and had learned to dance the Virginia Reel. . . . They had also made costumes with paper ruffles of Mexican colors to represent their flag. They prepared Mexican dances and songs.[24]

Despite such an early and valiant attempt at diversity, the teacher did not think it necessary to include the indigenous heritage of Mexican women. Indeed, stereotypical representations of the American Indian "princess" (or what Rayna Green has termed "the Pocahontas perplex"[25]) supplanted any understanding of indigenous cultures on either side of the political border separating Mexico and the United States.

Like Americanization advocates across the Southwest, Houchen settlement workers held out unrealistic notions of the American dream as well as romantic constructions of American life. It is as if the Houchen staff had endeavored to create a white, middle-class environment for Mexican youngsters complete with tutus and toe shoes. Cooking classes also became avenues for developing particular tastes. Minerva Franco, who as a child attended settlement programs and who later as an adult became a community volunteer, explained, "I'll never forget the look on my mother's face when I first cooked 'Eggs Benedict' which I learned to prepare at Houchen."[26] The following passage, taken from a report dated February 1942 outlines, in part, the perceived accomplishments of the settlement:

> Sanitary conditions have been improving—more children go to school—more parents are becoming citizens, more are leaving Catholicism—more are entering business and public life—and more and more they taking on the customs and standards of the Anglo people.[27]

Seemingly oblivious to structural discrimination, such a statement ignores economic segmentation and racial/ethnic segregation. Focusing on El Paso, historian Mario García demonstrated that the curricula in Mexican schools, which emphasized vocational education, served to funnel Mexican youth into the factories and building trades. In the abstract, education raised expectations, but in practice, particularly for men, it trained them for low-status, low-paying jobs. One California grower disdained education for Mexicans because it would give them "tastes for things they can't acquire."[28] Settlement workers seemed to ignore that racial/ethnic identity involved not only a matter of personal choice and heritage but also an ascribed status imposed by external sources.[29]

Americanization programs have come under a lot of criticisms from historians over the past two decades and numerous passages and photographs in the Houchen collection provide fodder for sarcasm among contemporary readers. Yet, to borrow from urban theorist Edward Soja, scholars should be mindful of "an appropriate interpretive balance between space, time and social being."[30] Although cringing at the ethnocentrism and romantic idealizations of "American" life, I respect the settlement workers for their health and child care services. Before judging the maternal missionaries too harshly, it is important to keep in mind the social services they rendered over an extended period of time as well as the environment in which they lived. For example, Houchen probably launched the first bilingual kindergarten program in El Paso, a program that eased the children's transition into an English-only first grade. Houchen residents did not denigrate the use of Spanish and many became fluent Spanish speakers. The hospital and clinic, moreover, were important community institutions for over half a century.[31]

Settlement workers themselves could not always count on the encouragement or patronage of Anglo El Paso. In a virulently nativist tract, a local physician, C. S. Babbitt, condemned missionaries, like the women of Houchen, for working among Mexican and African Americans. In fact, Babbitt argued that religious workers were "seemingly conspiring with Satan to destroy the handiwork of

God" because their energies were "wasted on beings . . . who are not in reality the objects of Christ's sacrifice."[32] Even within their own ranks, missionaries could not count on the support of Protestant clergy. Reverend Robert McLean, who worked among Mexicans in Los Angeles, referred to his congregation as "chili con carne" bound to give Uncle Sam a bad case of "heartburn."[33]

Perhaps more damaging than these racist pronouncements was the apparent lack of financial support on the part of El Paso area Methodist churches. Accessible records reveal little in terms of local donations. Houchen was named after a former Michigan schoolteacher who bequeathed $1,000 for the establishment of a settlement in El Paso. The Women's Home Missionary Society of the Newark, New Jersey, Conference proved instrumental in raising funds for the construction of both Freeman Clinic and Newark Methodist Maternity Hospital. When Freeman Clinic first opened its doors in June 1921, all the medical equipment—everything from sterilizers to baby scales—were gifts from Methodist groups across the nation. The Houchen Day Nursery, however, received consistent financial support from the El Paso Community Chest and later the United Way. In 1975, Houchen's Board of Directors conducted the first community-wide fund-raising drive. Volunteers sought to raise $375,000 to renovate existing structures and build a modern day care center. The Houchen fund-raising slogan "When people pay their own way, it's your affair . . . not welfare" makes painfully clear the conservative attitudes toward social welfare harbored by affluent El Pasoans.[34]

The women of Houchen appeared undaunted by the lack of local support. For over fifty years, these missionaries coordinated a multifaceted Americanization campaign among the residents of Segundo Barrio. How did Mexican women perceive the settlement? What services did they utilize and to what extent did they internalize the romantic notions of "Christian Americanization?"

Examining Mexican women's agency through institutional records is difficult; it involves getting beneath the text to dispel the shadows cast by missionary devotion to a simple Americanization ideology. One has to take into account the selectivity of voices. In drafting settlement reports and publications, missionaries chose those voices that would publicize their "victories" among the Spanish speaking. As a result, quotations abound that heap praise upon praise on Houchen and its staff. For example, in 1939, Soledad Burciaga emphatically declared, "There is not a person, no matter to which denomination they belong, who hasn't a kind word and a heart full of gratitude towards the Settlement House."[35] Obviously, these documents have their limits. Oral interviews and informal discussions with people who grew up in Segundo Barrio give a more balanced, less effusive perspective. Most viewed Houchen as a Protestant-run health care and after-school activities center rather than as the "light-house" [sic] in South El Paso.[36]

In 1949, the term Friendship Square was coined as a description for the settlement house, hospital, day nursery, and church. Missionaries hoped that children born at Newark would participate in preschool and afternoon programs and that eventually they and their families would join their church, El Buen Pastor. And a few did follow this pattern. One of the ministers assigned to El Buen Pastor, Fernando García, was a Houchen kindergarten graduate. Emulating the settlement staff, some young women enrolled in Methodist missionary colleges or served as lay volunteers. Elizabeth Soto, for example, attended Houchen programs throughout her childhood and adolescence. On graduation from Bowie High School, she entered Asbury College to train as a missionary and then returned to El Paso as a Houchen resident. After several years of service, she left settlement work to become the wife of a Mexican Methodist minister. The more common goal among Houchen teens was to graduate from high school and perhaps attend Texas Western, the local college. The first child born at Newark Hospital, Margaret Holguin, took part in settlement activities as a child and later became a registered nurse. According to her comadre, Lucy Lucero, Holguin's decision to pursue nursing was "perhaps due to the influence" of head nurse Millie Rickford. According to Lucero, "The only contact I had had with Anglos was with Anglo teachers. Then I met Miss Rickford and I felt, 'Hey, she's human. She's great.'" At a time when many (though certainly not all) elementary schoolteachers cared little about their Mexican students, Houchen residents offered warmth and encouragement.[37]

Emphasizing education among Mexican youth seemed a common goal characterizing Methodist community centers and schools. The Frances De Pauw School located on Sunset Boulevard in Los Angeles, for example, was an all-girls boarding school. Frances De Pauw educated approximately 1,800 young Mexican women from 1900 to 1946 and a Methodist pamphlet elaborated on its successes. "Among [the school's] graduates are secretaries, bookkeepers, clerks, office receptionists, nurses, teachers, waitresses, workers in cosmetic laboratories, church workers, and Christian homemakers." While preparing its charges for the workaday world, the school never lost sight of women's domestic duties. "Every De Pauw girl is graded as carefully in housework as she is in her studies."[38] With regard to Friendship Square, one cannot make wholesale generalizations about its role in fostering mobility or even aspirations for mobility among the youth of Segundo Barrio. Yet it is clear that Houchen missionaries strived to build self-esteem and encouraged young people to pursue higher education.

Missionaries also envisioned a Protestant enclave in South El Paso; but, to their frustration, very few people responded. The settlement church, El Buen Pastor, had a peak membership of 150 families. The church itself had an intermittent history. Shortly after its founding in 1897, El Buen Pastor disappeared; it was officially rededicated as part of Houchen in 1932. However, the construction of an actual church on settlement grounds did not begin until 1945. In 1968, the small rock chapel would be converted into a recreation room and thrift shop as the members of El Buen Pastor and El Mesias (another Mexican-American church) were merged together to form the congregation of the Emmanuel United Methodist Church in downtown El Paso. In 1991, a modern gymnasium occupies the ground where the chapel once stood.[39]

The case histories of converts suggest that many of those who joined El Buen Pastor were already Protestant. The Dominguez family offers an example. In the words of settlement worker A. Ruth Kern:

> Reyna and Gabriel Dominguez are Latin Americans, even though both were born in the United States. Some members of the family do not even speak English. Reyna was born . . . in a Catholic home, but at the age of eleven years, she began attending the Methodist Church. Gabriel was born in Arizona. His mother was a Catholic, but she became a Protestant when . . . Gabriel was five years old.[40]

The youth programs at Houchen brought Reyna and Gabriel together. After their marriage, the couple had six children, all born at Newark Hospital. The Dominguez family represented Friendship Square's typical success story. Many of the converts were children and many had already embraced a Protestant faith. In the records I examined, I found only one instance of the conversion of a Catholic adult and one of the conversion of an entire Catholic family.[41] It seems that those most receptive to Houchen's religious messages were already predisposed in that direction.

The failure of proselytization cannot be examined solely within the confines of Friendship Square. It is not as if these Methodist women were good social workers but incompetent missionaries. Houchen staff member Clara Sarmiento wrote of the difficulty in building trust among the adults of Segundo Barrio. "Though it is easy for children to open up their hearts to us we do not find it so with the parents." She continued, "It is hard especially because we are Protestant, and most of the people we serve . . . come from Catholic heritage."[42] I would argue that the Mexican community played an instrumental role in thwarting conversion. In a land where the barrio could serve as a refuge from prejudice and discrimination, the threat of social isolation could certainly inhibit many residents from turning Protestant. During an oral interview, Estella Ibarra, a woman who participated in Houchen activities for over fifty years, described growing up Protestant in South El Paso:

> We went through a lot of prejudice . . . sometimes my friends' mothers wouldn't let them play with us. . . . When the priest would go through the neighborhood, all the children would run to say hello and kiss his hand. My brothers and I would just stand by and look. The priest would usually come . . . and tell us how we were living in sin. Also, there were times when my brother and I were stoned by other students . . . and called bad names.[43]

When contacted by a Houchen resident, Mrs. Espinosa admitted to being a closet Protestant. As she explained, "I am afraid of the Catholic sisters and [I] don't want my neighbors to know that I am not Catholic-minded." The fear of ostracism, while recorded by Houchen staff, did not figure into their understanding of Mexicano resistance to conversion. Instead, they blamed time and culture. Or as Dorothy Little succinctly related, "We can not eradicate in a few years what has been built up during ages."[44] Their dilemma points to the fact historians Sarah Deutsch and George Sánchez have noted: Americanization programs in the Southwest, most of which were sporadic and poorly financed, made little headway in Mexican communities. Ruth Crocker also described the Protestant settlements in Gary, Indiana, as having only a "superficial and temporary" influence.[45] Yet even long-term sustained efforts, as in the case of Houchen, had limited appeal. This inability to mold consciousness or identity demonstrates not only the strength of community sanctions, but, more significant, of conscious decision-making on the part of Mexican women who sought to claim a place for themselves and their families in American society without abandoning their Mexican cultural affinities.

Mexican women derived substantive services from Friendship in the form of health care and education; however, they refused to embrace its romantic idealizations of American life. Wage-earning mothers who placed their children in the day nursery no doubt encountered an Anglo world quite different from the one depicted by Methodist missionaries and thus were skeptical of the settlement's cultural messages. Clara Sarmiento knew from experience that it was much easier to reach the children than their parents.[46] How did children respond to the ideological undercurrents of Houchen programs? Did Mexican women feel empowered by their interaction with the settlement or were Methodist missionaries invidious underminers of Mexican identity?

In getting beneath the text, the following remarks of Minerva Franco that appeared in a 1975 issue of *Newark-Houchen News* raise a series of provocative questions. "Houchen provided . . . opportunities for learning and experiencing. . . . At Houchen I was shown that I had worth and that I was an individual."[47] Now what did she mean by that statement? Did the settlement house heighten her self-esteem? Did she feel that she was not an individual within the context of her family and neighborhood? Some young women imbibed Americanization so heavily as to reject their identity. In *No Separate Refuge*, Sarah Deutsch picked up on this theme as she quoted missionary Polita Padilla: "I am Mexican, born and brought up in New Mexico, but much of my life was spent in the Allison School where we had a different training so that the Mexican way of living now seems strange to me." Others, like Estella Ibarra and Rose Escheverría Mulligan, saw little incompatibility between Mexican traditions and Protestantism.[48]

Which Mexican women embraced the ideas of assimilation so completely as to become closet Mexicans? As a factor, class must be taken into consideration. In his field notes housed at the Bancroft Library, economist Paul Taylor contends that middle-class Mexicans desiring to dissociate themselves from their working-class neighbors possessed the most fervent aspirations for assimilation. Once in the United States, middle-class Mexicanos found themselves subject to racial/ethnic prejudice that did not discriminate by class. Due to restrictive real estate covenants, immigrants lived in barrios with people they considered inferiors.[49] By passing as "Spanish," they cherished hopes of melting into the American social landscape. Sometimes mobility-minded parents sought to regulate their children's choice of friends and later marriage partners. "My folks never allowed us to around with Mexicans," remembered Alicia Mendeola Shelit. "We went sneaking around, but my Dad wouldn't allow it. We'd always be with white." Indeed, Shelit married twice, both times to Euro-Americans.[50] Of course it would be unfair to characterize all middle-class Mexican women immigrants as repudiating their mestizo identity. Working in a posh El Paso department store, Alma Araiza would quickly correct her colleagues when they assumed she was Italian.

> People kept telling me, 'You must not be Mexican.' And I said, 'why do you think I'm not?' 'Well, it's your skin color. Are you Italian?' . . . I [responded] 'I am Mexicana.'[51]

Or as a young woman cleverly remarked to anthropologist Ruth Tuck, "Listen, I may be a Mexican in a fur coat, but I'm still a Mexican."[52]

The Houchen documents reveal glimpses into the formation of identity, consciousness, and values. The Friendship Square Calendar of 1949 explicitly stated that the medical care provided at Houchen "is a tool to develop sound minds in sound bodies; for thus it is easier to find peace with God and man. We want to help people develop a sense of values in life." Furthermore, the privileging of color—with white as the pinnacle—was an early lesson. Relating the excitement of kindergarten graduation, Day Nursery head Beatrice Fernandez included in her report a question asked by Margarita, one of the young graduates. "We are all wearing white, white dress, slip, socks and Miss Fernandez, is it alright if our hair is black?"[53] Sometimes subtle, sometimes overt, the privileging of race, class, culture, and color taught by women missionaries had painful consequences for their pupils.

Houchen activities were synonymous with Americanization. A member of the settlement Brownie troop encouraged her friends "to become 'an American or a Girl Scout' at Houchen." Scouting certainly served as a vehicle for Americanization. The all-Mexican Girl and Boy Scout Troops of Alpine, Texas, enjoyed visiting El Paso and Ciudad Juárez in the company of Houchen scouts. In a thank-you note, the Alpine Girl Scouts wrote, "Now we can all say we have been to a foreign country."[54]

It is important to remember that Houchen provided a bilingual environment, not a bicultural one. Spanish was the means to communicate the message of Methodism and Christian Americanization. Whether dressing up children as Pilgrims or European peasants, missionaries stressed "American" citizenship and values; yet, outside conversion, definitions of those values or of "our Amer-

ican way" remained elusive. Indeed, some of the settlement lessons
were not incongruous with Mexican mores. In December 1952, a
Euro-American settlement worker recorded in her journal the success of a Girl Scout dinner. "The girls learned a lot from it too. They
were taught how to set the table, and how to serve the men. They
learned also that they had to share, to cooperate, and to wait their
turn."[55] These were not new lessons.

The most striking theme that repeatedly emerges from Houch-
en documents is that of individualism. Missionaries emphasized
the importance of individual decision-making and individual accomplishment. In recounting her own conversion, Clara Sarmiento
explained to a young client, "I chose my own religion because it was
my own personal experience and . . . I was glad my religion was not
chosen for me."[56]

In *Relations of Rescue,* Peggy Pascoe carefully recorded the
glass ceiling encountered by "native helpers" at Protestant rescue
homes. Chinese women at Cameron House in San Francisco, for
example, could only emulate Euro-American missionaries to a certain point, always as subordinates, not as directors or leaders. Conversely, Mexican women did assume top positions of leadership at
Methodist settlements. In 1930, María Moreno was appointed the
head resident of the brand new Floyd Street Settlement in Dallas,
Texas. Methodist community centers and boarding schools stressed the need for developing "Christian leaders trained for useful living."[57] For many, leadership meant ministering as a lay volunteer;
for some, it meant pursuing a missionary vocation.

The Latina missionaries of Houchen served as cultural brokers
as they diligently strived to integrate themselves into the community. Furthermore, over time Latinas appeared to have experienced
some mobility within the settlement hierarchy. In 1912, Ofilia [sic]
Chávez served as a "student helper"; forty years later Beatrice Fernandez would direct the preschool. Until 1950, the Houchen staff
usually included one Latina; however, during the 1950s, the number of Latina (predominately Mexican American) settlement workers rose to six. Mary Lou López, María Rico, Elizabeth Soto, Febe
Bonilla, Clara Sarmiento, María Payan, and Beatrice Fernandez
had participated in Methodist outreach activities as children (Soto
at Houchen) and had decided to follow in the footsteps of their
teachers. In addition, these women had the assistance of five full-
time Mexican laypersons.[58] It is no coincidence that the decade of
greatest change in Houchen policies occurred at a time when Latinas held a growing number of staff positions. Friendship Square's

greater sensitivity to neighborhood needs arose, in part, out of the influence exerted by Mexican clients in shaping the attitudes and actions of Mexican missionaries.

So, in the end, Mexican women utilized Houchen's social services; they did not, by and large, adopt its tenets of Christian Americanization. Children who attended settlement programs enjoyed the activities, but Friendship Square did not always leave a lasting imprint. "My Mom had an open mind, so I participated in a lot of clubs. But I didn't become Protestant," remarked Lucy Lucero. "I had fun and I learned a lot, too." Because of the warm, supportive environment, Houchen Settlement is remembered with fondness. However, one cannot equate pleasant memories with the acceptance of the settlement's cultural ideals.[59]

Settlement records bear out Mexican women's *selective* use of Houchen's resources. The most complete set of figures is for the year 1944. During this period, 7,614 people visited the clinic and hospital. The settlement afternoon programs had an average monthly enrollment of 362 and 40 children attended kindergarten. Taken together, approximately 8,000 residents of Segundo Barrio utilized Friendship Square's medical and educational offerings. In contrast, the congregation of El Buen Pastor included 160 people.[60] Although representing only a single year, these figures indicate the importance of Houchen's medical facilities and Mexican women's selective utilization of resources.

By the 1950s, settlement houses were few and far between and those that remained were run by professional social workers. Implemented by a growing Latina staff, client-initiated changes in Houchen policies brought a realistic recognition of the settlement as a social service agency rather than a religious mission. During the 1950s, brochures describing the day nursery emphasized that while children said grace at meals and sang Christian songs, they would not receive "in any way indoctrination" regarding Methodism. In fact, at the parents' request, Newark nurses summoned Catholic priests to the hospital to baptize premature infants. Client desire became the justification for allowing the presence of Catholic clergy, a policy that would have been unthinkable in the not too distant past.[61] Finally, in the new Houchen constitution of 1959, all mention of conversion was dropped. Instead, it conveyed a more ecumenical, nondenominational spirit. For instance, the goal of Houchen Settlement was henceforth "to establish a Christian democratic framework for—individual development, family solidarity, and neighborhood welfare."[62]

Settlement activities also became more closely linked with the

Mexican community. During the 1950s, Houchen was the home of two LULAC chapters—one for teenagers and one for adults. The League of United Latin American Citizens (LULAC) was the most visible and politically powerful civil rights organization in Texas.[63] Carpentry classes—once the preserve of males—opened their doors to young women, although on a gender-segregated basis. Houchen workers, moreover, made veiled references to the "very dangerous business" of Juárez abortion clinics; however, it appears unclear whether or not the residents themselves offered any contraceptive counseling. During the early 1960s, however, the settlement, in cooperation with Planned Parenthood, opened a birth control clinic for "married women." Indeed, a Houchen contraception success story was featured on the front page of a spring newsletter. "Mrs. G _____, after having her thirteenth and fourteenth children (twins), enrolled in our birth control clinic; now for one and one half years she has been a happy and non-pregnant mother."[64] Certainly Houchen had changed with the times. What factors accounted for the new directions in settlement work? The evidence on the baptism of premature babies seems fairly clear in terms of client pressure, but to what extent did other policies change as the result of Mexican women's input? The residents of Segundo Barrio may have felt more comfortable expressing their ideas and Latina settlement workers may have exhibited a greater willingness to listen. Indeed, Mexican clients, not missionaries, set the boundaries for interaction.

Creating the public space of settlements and community centers, advocates of Americanization sought to alter the "lifeworld" of Mexican immigrants to reflect their own idealized versions of life in the United States. Settlement workers can be viewed as the narrators of lived experience as Houchen records reflected the cognitive construction of missionary aspirations and expectations. In other words, the documents revealed more about the women who wrote them than those they served. At another level, one could interpret the cultural ideals of Americanization as an indication of an attempt at what Jürgen Habermas has termed "inner colonization."[65] Yet the failure of such projects illustrates the ways in which Mexican women appropriated desired resources, both material (infant immunizations) and psychological (self-esteem) while, in the main, rejecting the ideological messages behind them. The shift in Houchen policies during the 1950s meant more than a recognition of community needs; it represented a claiming of public space by Mexican women clients.

Confronting Americanization brings into sharp relief the concept I have termed cultural coalescence. Immigrants and their children pick, borrow, retain, and create distinctive cultural forms. There is no single hermetic Mexican or Mexican-American culture, but rather permeable *cultures* rooted in generation, gender, region, class, and personal experience. Chicano scholars have divided Mexican experiences into three generational categories: Mexicano (first generation), Mexican American (second generation), and Chicano (third and beyond).[66] But this general typology tends to obscure the ways in which people navigate across cultural boundaries as well as their conscious decision-making in the production of culture. However, people of color have not had unlimited choice. Race and gender prejudice and discrimination with their accompanying social, political, and economic segmentation have constrained aspirations, expectations, and decision-making.

The images and ideals of Americanization were a mixed lot and were never the only messages immigrant women received. Local *mutualistas*, Mexican patriotic and Catholic pageants, newspapers, and community networks reinforced familiar legacies. In contrast, religious and secular Americanization programs, the elementary schools, movies, magazines, and radio bombarded the Mexican community with a myriad of models, most of which were idealized, stylized, unrealistic, and unattainable. Expectations were raised in predictable ways. In the words of one Mexican-American woman, "We felt that if we worked hard, proved ourselves, we could become professional people."[67] Consumer culture would hit the barrio full force during the 1920s, exemplified by the Mexican flapper. As we will see in the next chapter, even Spanish-language newspapers promoted messages of consumption and acculturation. Settlement houses also mixed in popular entertainment with educational programs. According to historian Louise Año Nuevo Kerr, the Mexican Mothers Club of the University of Chicago Settlement "took a field trip to NBC radio studios in downtown Chicago from which many of the soap operas emanated."[68]

By looking through the lens of cultural coalescence, we can begin to discern the ways in which people select and create cultural forms. Teenagers began to manipulate and reshape the iconography of consumer culture both as a marker of peer group identity and as an authorial presence through which they rebelled against strict parental supervision. When standing at the cultural crossroads, Mexican women blended their options and created their own paths.

Figure 1 Giving the camera a glimpse of her steely determination, Pasquala Esparza, with her daughters Jesusita and baby Raquel, pose for a passport photo taken in Ciudad Juárez. Courtesy of Jesusita Torres.

Figure 2 Las solas: Mexican women arriving in El Paso, 1911. Courtesy of the Rio Grande Historical Collections, New Mexico State University Library, Las Cruces.

Figure 3 Mexican migrant women and children in front of company housing, Arizona 1930s. Courtesy of the Carey McWilliams Collection, Special Collections, University Research Library, UCLA.

Figure 4 Caring for a tubercular relative, Estella Córtez in the kitchen of the home she rented for four dollars per week in Corpus Christi, Texas, 1949. Courtesy of Lee (Russell) Photograph Collection, The Center for American History, The University of Texas at Austin, Neg. no. 13918-AF4.

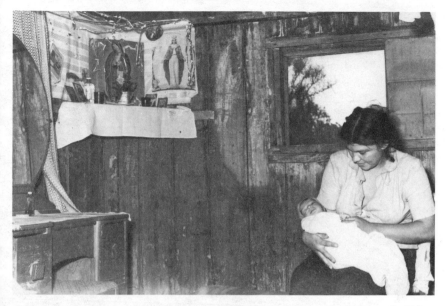

Figure 5 Mother and infant, 1949. Courtesy of Lee (Russell) Photograph Collection, The Center for American History, The University of Texas at Austin, Neg. no. 13918-EF12.

Figure 6 Identified only as a mother of eight, this Tejana was photographed as she sewed. Sewing for family members and for pennies as homeworkers, women made meager earnings stretch. Courtesy of Lee (Russell) Photograph Collection, The Center for American History, The University of Texas at Austin, Neg. no. 14233-55.

Figure 7 Father and daughter, 1949. Courtesy of Lee (Russell) Photograph Collection, The Center for American History, The University of Texas at Austin, Neg. no. 14233-2.

Figure 8 A young Mexican immigrant with her son in San Bernardino, c. 1926. Courtesy of José Rivas.

Figure 9 "Comadres" Teresa Grijalva de Orosco and Francisca Ocampo Quesada, 1912. Courtesy of the Ocampo Papers, Chicano Research Collection, Department of Archives and Manuscripts, Arizona State University, Tempe, Ariz.

Figure 10 Wedding photo of Macario and Guadalupe Hernández, Santa Paula, California, 1928. The couple recently celebrated their sixty-ninth anniversary. Courtesy of Esteban and Elia Hernández.

Religion played an important role in the lives of many Latinos in the Southwest.

Figure 11 To commemorate her first communion, Marina Briones poses for a portrait with her beaming madrina (godmother). Personal loan to the author.

Figure 12 Catholic women took great pride and comfort in their home altars. Courtesy of Lee (Russell) Photograph Collection, The Center for American History, The University of Texas at Austin, Neg. no. CN06311.

Figure 13 A Pentecostal congregation in Greeley, Colorado, 1932. This gathering is indicative of the small Protestant enclaves which endure in Spanish-speaking communities. Courtesy of Archives, City of Greeley Museums, Permanent Collection.

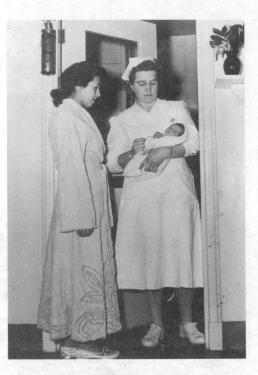

Figure 14 From 1937 to 1976, over 12,000 babies were born at Newark Hospital adjacent to the Houchen Settlement. Although mother and child are unidentified, the nurse is Dorothea Muñoz. Courtesy of Houchen Community Center, El Paso, Tex.

Figure 15 For many years, the big attraction for children of El Paso's Segundo Barrio was the elaborate playground at Houchen Settlement. Courtesy of Houchen Community Center, El Paso, Tex.

Figure 16 "Christian Americanization" guided all of Houchen's programs. Here kindergarten students say grace during snack-time. Courtesy of Houchen Community Center, El Paso, Tex.

Figure 17 For fifty cents per week, children could take dance or music classes and then perform in Houchen recitals. Courtesy of Houchen Community Center, El Paso, Tex.

Figure 18 Emblematic of cultural coalescence, María Soto Audelo at a Fourth of July celebration, 1917. Courtesy of the Arizona Historical Society, Tucson.

Figure 19 The great-aunt of singer Linda Ronstadt, entertainer Luisa Espinel in elegant flapper attire. Courtesy of the Arizona Historical Society Library, Tucson.

Figure 20 Fashion lay-outs, celebrity testimonials, and cosmetic ads promoted the
iconographies of Americanization and consumption to a Spanish-speaking public. *La
Opinión,* May 16, 1927, reprinted with
permission.

MODELOS DE NUEVA YORK

Hay varios detalles que
concurren en la hechura de
este abrigo de Verano que lo
hacen elegante. El cuello, por
ejemplo, presenta un efecto
nuevo, lo mismo que las man
gas y la combinación de la te
la negra de crepé de seda con
la piel de zorra.

La aplicación de pieles en
la parte inferior de los abri-
gos, constituye lo más nove-
doso, máxime cuando la combinación se hace con buen
gusto.

Para completar el conjunto,
se puede llevar un sombrero
forrado de seda y unas zapati
llas color negro, de cabriti-
lla.

Figure 21 Note the types of products designed to "whiten" one's complexion. *La Opinión*, June 5, 1927, reprinted with permission.

Figure 22 Promises of status and affection were the staples of cosmetics advertising in both English and Spanish. *La Opinión*, February 8, 1938, reprinted with permission.

Figure 23 My mother as a young woman, Erminia Ruiz, 1941.
Personal collection of the author.

Figure 24 Big Band Music with a Latin Beat. The Quintero's Orchestra, 1947. Courtesy of the Arizona Historical Society, Tucson.

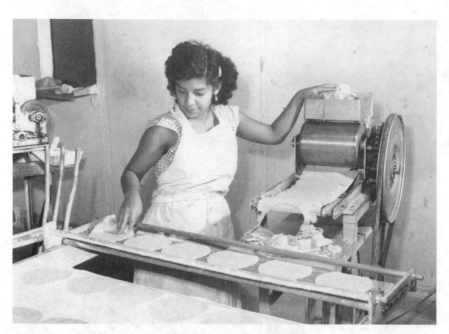

Figure 25 An employee of the La Malinche Tortilla Factory, Corpus Christi, Texas, 1949. Courtesy of Lee (Russell) Photograph Collection, The Center for American History, The University of Texas at Austin, Neg. no. II DF-6.

Figure 26 For Mexican-American women, clerical jobs represented status and mobility. Courtesy of Lee (Russell) Photograph Collection, The Center for American History, The University of Texas at Austin, Neg. no. II DF-10.

3

The Flapper and the Chaperone

❖

IMAGINE a gathering in a barrio hall, a group of young people dressed "to the nines" trying their best to replicate the dance steps of Fred Astaire and Ginger Rogers. This convivial heterosocial scene was a typical one in the lives of teenagers during the interwar period. But along the walls, a sharp difference was apparent in the barrios. Mothers, fathers, and older relatives chatted with one another as they kept one eye trained on the dance floor. They were the chaperones—the ubiquitous companions of unmarried Mexican-American women. Chaperonage was a traditional instrument of social control. Indeed, the presence of *la dueña* was the prerequisite for attendance at a dance, a movie, or even church-related events. "When we would go to town, I would want to say something to a guy. I couldn't because my mother was always there," remembered María Ybarra. "She would always stick to us girls like glue. . . . She never let us out of her sight."[1]

An examination of events like this one reveals the ways in which young Mexican women in the United States between the wars rationalized, resisted, and evaded parental supervision. It offers a glimpse into generational conflict that goes beyond the more general differences in acculturation between immigrants and their children. Chaperonage existed for centuries on both sides of the political border separating Mexico and the United States. While conjuring images of patriarchal domination, chaperonage is best understood as a manifestation of familial oligarchy whereby elders attempted to dictate the activities of youth for the sake of family honor. A family's standing in the community depended, in part, on women's purity. Loss of virginity not only tainted the reputation of an individual, but of her kin as well. For Mexicano immigrants living in a new, bewildering environment filled with temptations, the enforcement of chaperonage assumed a particular urgency.[2]

Historians Donna Gabaccia and Sydney Stahl Weinberg have urged immigration historians to notice the subtle ways women shaped and reshaped their environments, especially within the family. In addition, pathbreaking works by Elizabeth Ewen, Andrew Heinze, and Susan Glenn examine the impact of U.S. consumer culture on European immigrants.[3] Indeed, while some Mexican-American youth negotiated missionary idealizations of American life, other teenagers sought the American dream as promised in magazines, movies, and radio programs.

Popular culture could serve as a tool for literacy. While mothers may have viewed romance magazines, like *True Story*, as giving their daughters "bad ideas,"[4] the following passage taken from my interview with Jesusita Torres poignantly reveals a discourse of discovery:

> I started reading in English. I remember when I started reading the love story books . . . I started understanding. But I never knew the end of it. You know why I never knew the end of the story? Because . . . when you turn the page and at the end it says 'cont.' like in continue, I never knew what it meant. So one day I discovered it myself. I was reading and I always wanted to know the end of the story. I wanted to know so bad so I continued flipping the pages, then I looked . . . the same name of the story and it said the same word 'cont' . . . I learned that. And then from then on . . . I finished the stories. . . . To me that was something . . . you see, I did not have anybody to say: "Look, you look at the page and you continue."[5]

The transmitters of Americanization, including mass culture, could influence women in many ways—from fantasy to lived experience, from hegemonic presence to small, personal victories. I am reminded of the words of George Lipsitz: "Images and icons compete for dominance within a multiplicity of discourses." He continues, "Consumers of popular culture move in and out of subject positions in a way that allows the same message to have widely varying meanings at the point of reception." Or as Indonesian poet Nirwan

Dewanto stated, "The struggle for our survival in the information era is not to be won where information originates, but where it arrives."[6]

Confronting "America" began at an early age. Throughout the Southwest, Spanish-speaking children had to sink or swim in an English-only environment. Even on the playground, students were punished for conversing in Spanish. Admonishments, such as "Don't speak that ugly language, you are an American now," not only reflected a strong belief in Anglo conformity but denigrated the self-esteem of Mexican-American children.[7] As Mary Luna stated:

> It was rough because I didn't know English. The teacher wouldn't let us talk Spanish. How can you talk to anybody? If you can't talk Spanish and you can't talk English. . . . It wasn't until maybe the fourth or fifth grade that I started catching up. And all that time I just felt I was stupid.[8]

Yet Luna credited her love of reading to a Euro-American educator who had converted a small barrio house into a makeshift community center and library. Her words underscore the dual thrust of Americanization—education and consumerism. "To this day I just love going into libraries . . . there are two places that I can go in and get a real warm, happy feeling; that is, the library and Bullock's in the perfume and make-up department."[9]

Racial/ethnic women struggled with boundaries and made decisions as individuals, family members, neighbors, and peers. The following works in Asian-American history seem especially enlightening. Taking into account a buffet of cultural choices emanating from Japanese communities and U.S. society at large, Valerie Matsumoto's study of Nisei women teenagers before the Second World War elegantly outlines the social construction of the Nisei world. Matsumoto emphasizes the agency of adolescents in creating and nurturing their own youth culture. Along similar lines, Judy Yung in *Unbound Feet* beautifully details the work and leisure activities among Chinese-American teenagers in San Francisco.[10]

For Mexican Americans, second-generation women as teenagers have received scant scholarly attention. Among Chicano historians and writers, there appears a fascination with the sons of immigrants, especially as *pachucos*.[11] Young women, however, may have experienced deeper generational tensions as they blended elements of Americanization with Mexican expectations and values. This chapter focuses on the shifting interplay of gender, cultures, class, ethnicity, and youth and the ways in which women negotiate across specific cultural contexts blending elements as diverse as celebrating Cinco de Mayo and applying Max Factor cosmetics.

In grappling with Mexican-American women's consciousness and agency, oral history offers a venue for exploring teenage expectations and preserving a historical memory of attitudes and feelings. In addition to archival research, the recollections of seventeen women serve as the basis for my reconstruction of adolescent aspirations and experiences (or dreams and routines).[12] The women themselves are fairly homogeneous in terms of nativity, class, residence, and family structure. With two exceptions, they are U.S. citizens by birth and attended southwestern schools. All the interviewees were born between 1908 and 1926.[13] Although three came from families once considered middle class in Mexico, most can be considered working class in the United States. Their fathers' typical occupations included farm worker, miner, day laborer, and railroad hand. These women usually characterized their mothers as homemakers, although several remembered that their mothers took seasonal jobs in area factories and fields. The most economically privileged woman in the sample, Ruby Estrada, helped out in her family-owned hardware and furniture store. She is also the only interviewee who attended college.[14] It should be noted that seven of the seventeen narrators married Euro-Americans. Although intermarriage was uncommon, these oral histories give us insight into the lives of those who negotiated across cultures in a deeply personal way and who felt the impact of acculturation most keenly. Rich in emotion and detail, these interviews reveal women's conscious decision-making in the production of culture. In creating their own cultural spaces, the interwar generation challenged the trappings of familial oligarchy.

Chicano social scientists have generally portrayed women as "the 'glue' that keeps the Chicano family together" as well as the guardians of traditional culture.[15] Whether one accepts this premise or not, within families, young women, perhaps more than their brothers, were expected to uphold certain standards. Parents, therefore, often assumed what they perceived as their unquestionable prerogative to regulate the actions and attitudes of their adolescent daughters. Teenagers, on the other hand, did not always acquiesce in the boundaries set down for them by their elders. Intergenerational tension flared along several fronts.

Like U.S. teenagers, in general, the first area of disagreement between an adolescent and her family would be over her personal appearance. As reflected in F. Scott Fitzgerald's "Bernice Bobs Her Hair," the length of a young woman's tresses was a hot issue spanning class, region, and ethnic lines. During the 1920s, a woman's decision "to bob or not bob" her hair assumed classic proportions within Mexican families. After considerable pleading, Belen Martínez Mason was permitted to cut her hair, though she soon regretted the decision. "Oh, I cried for a month."[16] Differing opinions over fashions often caused ill feelings. One Mexican American woman recalled that as a young girl, her mother dressed her "like a nun" and she could wear "no make-up, no cream, no nothing" on her face. Swimwear, bloomers, and short skirts also became sources of controversy. Some teenagers left home in one outfit and changed into another at school. Once María Fierro arrived home in her bloomers. Her father inquired, "Where have you been dressed like that, like a clown?" "I told him the truth," Fierro explained "He whipped me anyway. . . . So from then on whenever I went to the track meet, I used to change my bloomers so that he wouldn't see that I had gone again."[17] The impact of flapper styles on the Mexican community was clearly expressed in the following verse taken from a corrido appropriately entitled "Las Pelonas" [The Bobbed-Haired Girls]:

> Red Banannas [sic]
> I detest,
> And now the flappers
> Use them for their dress.
> The girls of San Antonio
> Are lazy at the *metate*.
> They want to walk out bobbed-haired,
> With straw hats on.
> The harvesting is finished,
> So is the cotton;
> The flappers stroll out now
> For a good time.[18]

With similar sarcasm, another popular ballad chastised Mexican women for applying makeup so heavily as to resemble a piñata.[19]

The use of cosmetics, however, cannot be blamed entirely on Madison Avenue ad campaigns. The innumerable barrio beauty pageants, sponsored by *mutualistas*, patriotic societies, churches, the Mexican Chamber of Commerce, newspapers, and even progressive labor unions, encouraged young women to accentuate their physical attributes. Carefully chaperoned, many teenagers did participate in community contests from La Reina de Cinco de Mayo to Orange Queen. They modeled evening gowns, rode on parade floats, and sold raffle tickets.[20] Carmen Bernal Escobar remembered one incident where, as a contestant, she had to sell raffle tickets. Every ticket she sold counted as a vote for her in the pageant. Naturally the winner would be the woman who had accumulated the most votes. When her brother offered to buy $25 worth of votes [her mother would not think of letting her peddle the tickets at work or in the neighborhood], Escobar, on a pragmatic note, asked him to give her the money so that she could buy a coat she had spotted while window-shopping.[21]

The commercialization of personal grooming made additional inroads into the Mexican community with the appearance of barrio beauty parlors. Working as a beautician conferred a certain degree of status—"a nice, clean job"—in comparison to factory or domestic work. As one woman related:

> I always wanted to be a beauty operator. I loved makeup; I loved to dress up and fix up. I used to set my sisters' hair. So I had that in the back of my mind for a long time, and my mom pushed the fact that she wanted me to have a profession—seeing that I wasn't thinking of getting married.[22]

While further research is needed, one can speculate that neighborhood beauty shops reinforced women's networks and became places where they could relax, exchange *chimse* (gossip), and enjoy the company of other women.

During the 1920s, the ethic of consumption became inextricably linked to making it in America.[23] The message of affluence attainable through hard work and a bit of luck was reinforced in English and Spanish-language publications. Mexican barrios were not immune from the burgeoning consumer culture. The society pages of the influential Los Angeles-based *La Opinion*, for example, featured advice columns, horoscopes, and celebrity gossip. Advertisements for makeup, clothing, even feminine hygiene products reminded teenagers of an awaiting world of consumption.[24] One week after its inaugural issue in 1926, *La Opinion* featured a Spanish translation of Louella Parsons' nationally syndicated gossip column. Advertisements not only hawked products but offered instructions for behavior. As historian Roberto Treviño related in his recent study of Tejano newspapers, "The point remains that the

Spanish-language press conveyed symbolic American norms and models to a potentially assimilable readership."[25]

Advertisements aimed at women promised status and affection if the proper bleaching cream, hair coloring, and cosmetics were purchased. Or, as one company boldly claimed, "Those with lighter, more healthy skin tones will become much more successful in business love, and society."[26] A print ad [in English] for Camay Soap carried by *Hispano America* in 1932 reminded women readers that "Life Is a Beauty Contest."[27] Flapper fashions and celebrity testimonials further fused the connections between gendered identity and consumer culture. Another promotion encouraged readers to "SIGA LAS ESTRELLAS" (FOLLOW THE STARS) and use Max Factor cosmetics. It is important to keep in mind that Spanish-language newspapers filtered to their readers not only the iconography of U.S. popular culture, but also their perceptions of gender relations within that culture. For example, an advertisement for Godefroy's "Larieuse" hair coloring featured an attractive woman in profile smiling at the tiny man cupped in the palm of her hand. The diminutive male figure is shown on bended knee with his hands outstretched in total adoration. Does this hair coloring promotion found in the February 8, 1938, issue of *La Opinion* relay the impression that by using this Anglo product Mexican women will exert the same degree of power over their men as their Anglo peers supposedly plied?[28]

These visual representations raise all sorts of speculation as to their meaning, specifically with regard to the social construction of gender. I cannot identify the designers of these layouts, but the architects are less important than the subtle and not-so-subtle messages codified within their text. Mexican women interpreted these visual representations in a myriad of ways. Some ignored them, some redefined their messages, and other internalized them. The popularity of bleaching creams offers a poignant testament to color consciousness in Mexican communities, a historical consciousness accentuated by Americanization through education and popular culture.[29]

Reflecting the coalescence of Mexican and U.S. cultures, Spanish-language publications promoted pride in Latino theater and music while at the same time celebrated the icons of Americanization and consumption. Because of its proximity to Hollywood, *La Opinion* ran contests in which the lucky winner would receive a screen test. On the one hand, *La Opinion* nurtured the dreams of "success" through entertainment and consumption while, on the other, the newspaper railed against the deportations and repatriations of the 1930s.[30] Sparked by manufactured fantasies and clinging to youthful hopes, many Mexican women teenagers avidly read celebrity gossip columns, attended Saturday matinees, cruised Hollywood and Vine, and nurtured their visions of stardom. A handful of Latina actresses, especially Dolores del Rio and Lupe Velez, whetted these aspirations and served as public role models of the "American dream." As a *La Opinion* article on Lupe Velez idealistically claimed, "Art has neither nationalities nor borders."[31]

In her essay "City Lights: Immigrant Women and the Rise of the Movies," Elizabeth Ewen has argued that during the early decades of the twentieth century, "The social authority of the media of mass culture replaced older forms of family authority and behavior." Ewen further explained that the "authority of this new culture organized itself around the premise of freedom from customary bonds as a way of turning people's attention to the consumer market place as a source of self-definition."[32] Yet Mexican women had choices (though certainly circumscribed by economic considerations) about what elements to embrace and which to ignore. As George Lipsitz reminds us in *Time Passages*, "Hegemony is not just imposed on society from the top; it is struggled for from below, and no terrain is a more important part of that struggle than popular culture."[33] Mexican-American women teenagers also positioned themselves within the cultural messages they gleaned from English and Spanish-language publications, afternoon matinees, and popular radio programs. Their shifting conceptions of acceptable heterosocial behavior, including their desire "to date," heightened existing generational tensions between parents and daughters.[34]

Obviously, the most serious point of contention between an adolescent daughter and her Mexican parents regarded her behavior toward young men. In both cities and rural towns, close chaperonage was a way of life. Recalling the supervisory role played by her "old maid" aunt, María Fierro laughingly explained, "She'd check up on us all the time. I used to get so mad at her." Ruby Estrada recalled that in her small southern Arizona community, "all the mothers" escorted their daughters to the local dances. Estrada's mother was no exception when it came to chaperoning her daughters. "She went especially for us. She'd just sit there and take care of our coats and watch us." Even talking to male peers in broad daylight could be grounds for discipline.[35] Adele Hernández Milligan, a resident of Los Angeles for over fifty years, elaborated:

> I remember the first time that I walked home with a boy from school. Anyway, my mother saw me and she was mad. I must have been sixteen or seventeen. She slapped my face because I was walking home with a boy.[36]

Describing this familial protectiveness, one social scientist remarked that the "supervision of the Mexican parent is so strict as to be obnoxious."[37]

Faced with this type of situation, young women had three options: they could accept the rules set down for them; they could rebel; or they could find ways to compromise or circumvent traditional standards. "I was *never* allowed to go out by myself in the evening; it just was not done," related Carmen Bernal Escobar. In rural communities, where restrictions were perhaps even more stringent, "nice" teenagers could not even swim with male peers. According to Ruby Estrada, "We were ladies and wouldn't go swimming out there with a bunch of boys." Yet many seemed to accept these limits with equanimity. Remembering her mother as her chaperone, Lucy Acosta insisted, "I could care less as long I danced." "It wasn't devastating at all," echoed Ruby Estrada. "We took it in stride. We never thought of it as cruel or mean. . . . It was taken for granted that that's the way it was."[38] In Sonora, Arizona, like other small towns, relatives and neighbors kept close watch over adolescent women and quickly reported any suspected indiscretions. "They were always spying on you," Estrada remarked. Women in cities had a distinct advantage over their rural peers in that they could venture miles from their neighborhood into the anonymity of dance halls, amusement parks, and other forms of commercialized leisure. With carnival rides and the Cinderella Ballroom, the Nu-Pike amusement park of Long Beach proved a popular hangout for Mexican youth in Los Angeles.[39] It was more difficult to abide by traditional norms when excitement loomed just on the other side of the streetcar line.

Some women openly rebelled. They moved out of their family homes and into apartments. Considering themselves freewheeling single women, they could go out with men unsupervised as was the practice among their Anglo peers. Others challenged parental and cultural standards even further by living with their boyfriends. In his field notes, University of California economist Paul Taylor recorded an incident in which a young woman had moved in with her Anglo boyfriend after he had convinced her that such arrangements were common among Americans. "This terrible freedom in the United States," one Mexicana lamented. "I do not have to worry because I have no daughters, but the poor *señoras* with many girls, they worry."[40]

Those teenagers who did not wish to defy their parents openly would "sneak out" of the house to meet their dates or attend dances with female friends. Whether meeting someone at a drugstore, roller rink, or theater, this practice involved the invention of elaborate stories to mask traditionally inappropriate behavior.[41] In other words, they lied. In his study of Tuscon's Mexican community, Thomas Sheridan related the following saga of Jacinta Pérez de Valdez:

> As she and her sisters grew older, they used to sneak out of the house to go to the Riverside Ball Room. One time a friend of their father saw them there and said, "Listen, Felipe, don't you know your daughters are hanging around the Riverside?" Furious, their father threw a coat over his longjohns and stormed into the dance hall, not even stopping to tie his shoes. . . . Doña Jacinta recalled. "He entered by one door and we left by another. We had to walk back home along the railroad tracks in our high heels. I think we left those heels on the rails." She added that when their father returned, "We were all lying in bed like little angels."[42]

A more subtle form of rebellion was early marriage. By marrying at fifteen or sixteen, these women sought to escape parental supervision; yet it could be argued that, for many of these child brides, they exchanged one form of supervision for another in addition to the responsibilities of child-rearing.[43] In her 1933 ethnography, Clara Smith related the gripping testimony of one teenage bride:

> You see, my father and mother wouldn't let us get married. . . . Mother made me stay with her all the time. She always goes to church every morning at seven-thirty as she did in Mexico. I said I was sick. She went with my brothers and we just ran away and got married at the court. . . . They were strict with my sister, too. That's why she took poison and died.[44]

One can only speculate on the psychic pressures and external circumstances that would drive a young woman to take her own life.

Elopement occurred frequently since many parents believed

that no one was good enough for their daughters. "I didn't want to elope . . . so this was the next best thing to a wedding," recalled María Ybarra as she described how the justice of the peace performed the ceremony in her parents' home. "Neither my Dad or my Mom liked my husband. Nobody liked him," she continued. "My husband used to run around a lot. After we got married, he did settle down, but my parents didn't know that then."[45] One fifteen year old locked her grandmother in the outhouse so she could elope with her boyfriend. Indeed, when he first approached her at a San Joaquin Valley migrant camp asking if he could be her *novio*, she supposedly replied, "No, but I'll marry you." Lupe was just that desperate to escape familial supervision.[46]

If acquiescence, apartment living, early marriage, or elopement were out of the question, what other tactics did teenagers devise? The third alternative sometimes involved quite a bit of creativity on the part of young women as they sought to circumvent traditional chaperonage. Alicia Mendeola Shelit recalled that one of her older brothers would accompany her to dances ostensibly as a chaperone. "But then my oldest brother would always have a blind date for me." Carmen Bernal Escobar was permitted to entertain her boyfriends at home, but only under the supervision of her brother or mother. The practice of "going out with the girls," though not accepted until the 1940s, was fairly common. Several Mexican-American women, often related, would escort one another to an event (such as a dance), socialize with the men in attendance, and then walk home together. In the sample of seventeen interviews, daughters negotiated their activities with their parents. Older siblings and extended kin appeared in the background as either chaperones or accomplices. Although unwed teenage mothers were not unknown in Mexican barrios, families expected adolescent women to conform to strict standards of behavior.[47]

As can be expected, many teenage women knew little about sex other than what they picked up from friends, romance magazines, and the local theater. As Mary Luna remembered, "I thought that if somebody kissed you, you could get pregnant." In *Singing for My Echo,* New Mexico native Gregorita Rodríguez confided that on her wedding night, she knelt down and said her rosary until her husband gently asked, "Gregorita, *mi esposa,* are you afraid of me?" At times this naiveté persisted beyond the wedding. "It took four days for my husband to touch me," one woman revealed. "I slept with dress and all. We were both greenhorns, I guess."[48]

Of course, some young women did lead more adventurous

lives. A male interviewer employed by Mexican anthropologist
Manuel Gamio recalled his "relations" with a woman met at a Los
Angeles dance hall. Although born in Hermosillo, Elisa "Elsie"
Morales considered herself Spanish. She helped support her family by dancing with strangers. Even though she lived at home and
her mother and brother attempted to monitor her actions, she
managed to meet the interviewer at a "hot pillow" hotel. To prevent
pregnancy, she relied on contraceptive douches provided by "an
American doctor." Although Morales realized her mother would
not approve of her behavior, she noted that "she [her mother] is
from Mexico . . . I am from there also but I was brought up in the
United States, we think about things differently." Just as Morales
rationalized her actions as "American," the interviewer perceived
her within a similar, though certainly less favorable, definition of
Americanization. "She seemed very coarse to me. That is, she dealt
with one in the American way." Popular corridos, such as "El Enganchado" and "Las Pelonas," also touched on the theme of the
corrupting influence of U.S. ways on Mexican women.[49] If there
were rewards for women who escaped parental boundaries, there
were also sanctions for those who crossed established lines.[50]

Women who had children out of wedlock seemed to be treated
by their parents in one of two ways—as pariahs or prodigal daughters. Erminia Ruiz recalled the experiences of two girlhood friends:

> It was a disgrace to the whole family. The whole family suffered
> and . . . her mother said she didn't want her home. She could not
> bring the baby home and she was not welcome at home. . . . She
> had no place to go. . . . And then I had another friend. She was
> also pregnant and the mother actually went to court to try to get
> him to marry her. . . . He hurried and married someone else but
> then he had to give child support.[51]

In another instance, Carmen and her baby were accepted by
her family. She was, however, expected to work in the fields to support her infant. Her parents kept a watchful eye on her activities.
When Diego, a young Mexicano immigrant, asked Carmen out to
dinner, her baby became her chaperone. "[My mother] said to take
the baby with you. She was so smart so I wouldn't go any farther
than the restaurant. At first, I was ashamed but [he] said, "Bring
him so he can eat." Before Carmen accepted Diego's proposal of
marriage, she asked him for his family's address in Mexico so she
could make sure he was not already married. "I told him I wouldn't
answer him until I got an answer." Diego's mother replied that her

son was single, but had a girlfriend waiting for him. Carmen and Diego have been married for over fifty years.[52]

Carmen's story illustrates the resiliency and resourcefulness of Mexican-American women. Her behavior during her courtship with Diego demonstrates shrewdness and independence. Once burned, she would be nobody's *pendeja* (fool).

Yet autonomy on the part of young women was hard to win in a world where pregnant, unmarried teenagers served as community "examples" of what might happen to you or your daughter if appropriate measures were not taken. As an elderly Mexicana remarked, "Your reputation was everything."[53] In this sense, the chaperone not only protected the young woman's position in the community, but that of the entire family.

Chaperonage thus exacerbated conflict not only between generations but within individuals as well. In gaily recounting tales of ditching the *dueña* or sneaking down the stairwell, the laughter of the interviewees fails to hide the painful memories of breaking away from familial expectations. Their words resonate with the dilemma of reconciling their search for autonomy with their desire for parental affirmation. It is important to note that every informant who challenged or circumvented chaperonage held a full-time job, as either a factory or service worker. In contrast, most woman who accepted constant supervision did not work for wages. Perhaps because they labored for long hours, for little pay, and frequently under hazardous conditions, factory and service workers were determined to exercise some control over their leisure time. Indeed, Douglas Monroy has argued that outside employment "facilitated greater freedom of activity and more assertiveness in the family for Mexicanas."[54]

It may also be significant that none of the employed teenagers had attended high school. They entered the labor market directly after or even before the completion of the eighth grade. Like many female factory workers in the United States, most Mexican operatives were young, unmarried daughters whose wage labor was essential to the economic survival of their families. As members of a "family wage economy," they relinquished all or part of their wages to their elders. According to a 1933 University of California study, of the Mexican families surveyed with working children, the children's monetary contributions constituted 35 percent of total household income.[55] Cognizant of their earning power, they resented the lack of personal autonomy.

Delicate negotiations ensued as both parents and daughters

struggled over questions of leisure activities and discretionary income. Could a young woman retain a portion of her wages for her own use? If elders demanded every penny, daughters might be more inclined to splurge on a new outfit or other personal item on their way home from work or, even more extreme, they might choose to move out, taking their paychecks with them. Recognizing their dependence on their children's income, some parents compromised. Their concessions, however, generally took the form of allocating spending money rather than relaxing traditional supervision. Still, women's earning power could be an important bargaining chip.[56]

On one level, many teenagers were devoted to their parents as evident (at least, in part) by their employment in hazardous, low-paying jobs. For example, Julia Luna Mount recalled her first day at a Los Angeles cannery:

> I didn't have money for gloves so I peeled chiles all day long by hand. After work, my hands were red, swollen, and I was on fire! On the streetcar going home, I could hardly hold on my hands hurt so much. The minute I got home, I soaked my hands in a pan of cold water. My father saw how I was suffering and he said, *"Mi hija,* you don't have to go back there tomorrow," and I didn't.[57]

On the other hand, adolescents rebelled against what they perceived as an embarrassingly old-fashioned intrusion into their private lives. When chastised by her aunt for dancing too close to her partner, Alma Araiza García would retort, "I am not going to get pregnant just by leaning on his cheek, okay?"[58] They wanted the right to choose their own companions and to use their own judgment.

Chaperonage triggered deep-seated tensions over autonomy and self-determination. "Whose life is it anyway?" was a recurring question with no satisfactory answer. Many women wanted their parents to consider them dutiful daughters, but they also desired degrees of freedom. While ethnographies provide scintillating tales of teenage rebellion, the voices of the interviewees do not. Their stories reflect the experiences of those adolescents who struggled with boundaries. How can one retain one's "good name" while experiencing the joys of youth? How can one be both a good daughter and an independent woman?

To complete the picture, we also have to consider the perspective of Mexican immigrant parents who encountered a youth culture very different from that of their generation. For them, courtship had occurred in the plaza; young women and men promenaded under the watchful eyes of town elders, an atmosphere in which an exchange of meaningful glances could well portend engagement. One can understand their consternation as they watched their daughters apply cosmetics and adopt the apparel advertised in fashion magazines. In other words, "If she dresses like a flapper, will she then act like one?" Seeds of suspicion reaffirmed the penchant for traditional supervision.

Parents could not completely cloister their children from the temptations of "modern" society, but chaperonage provided a way of monitoring their activities. It was an attempt to mold young women into sheltered young matrons. But one cannot regard the presence of *la dueña* as simply an old world tradition on a collision course with twentieth-century life. The regulation of daughters involved more than a conflict between peasant ways and modern ideas. Chaperonage was both an actual and symbolic assertion of familial oligarchy. A family's reputation was linked to the purity of women. As reiterated in a Catholic catechism, if a young woman became a "faded lily," she and her family would suffer dire consequences.[59] Since family honor rested, to some degree, on the preservation of female chastity (or *vergüenza*), women were to be controlled for the collective good, with older relatives assuming unquestioned responsibility in this regard. Mexican women coming of age during the 1920s and 1930s were not the first to challenge the authority of elders. Ramón Gutiérrez in his pathbreaking scholarship on colonial New Mexico uncovered numerous instances of women who tried to exercise some autonomy over their sexuality.[60] The Mexican-American generation, however, had a potent ally unavailable to their foremothers—consumer culture.

United States consumerism did not bring about the disintegration of familial oligarchy, but it did serve as a catalyst for change. The ideology of control was shaken by consumer culture and the heterosocial world of urban youth. As previously indicated, chaperonage proved much easier to enforce in a small town. Ruby Estrada described how a young woman would get the third degree if caught with a potential boyfriend alone. "And they [the elders] would say what are [you] doing there all alone. . . . Yeah, what were you up to or if you weren't up to no good, why should you be talking to that boy?"[61]

In contrast, parents in the barrios of major cities fought a losing battle against urban anonymity and commercialized leisure. The Catholic Church was quick to point out the "dangerous amusement" inherent in dancing, theater-going, dressing fashionably, and reading pulp fiction. Under the section, "The Enemy in the Ballroom," a Catholic advice book warned of the hidden temptations of dance. "I know that some persons can indulge in it without harm; but sometimes even the coldest temperaments are heated by it."[62] Therefore, the author offered the following rules:

> (1) If you know nothing at all . . . about dancing do not trouble yourself to learn (2) Be watchful . . . and see that your pleasure in dancing does not grow into a passion. . . . (3) Never frequent fairs, picnics, carnivals, or public dancing halls where Heaven only knows what sorts of people congregate. (4) Dance only at private parties where your father or mother is present.[63]

Pious pronouncements such as these had little impact on those adolescents who cherished the opportunity to look and act like vamps and flappers.

Attempting to regulate the social life of young parishioners, barrio priests organized gender-segregated teen groups. In Los Angeles, Juventud Católica Feminina Mexicana (JCFM) had over fifty chapters. In her autobiography *Hoyt Street,* Mary Helen Ponce remembered the group as one organized for "nice" girls with the navy blue uniform as its most appealing feature. The local chapter fell apart during World War II as young women rushed off to do their patriotic duty at "canteens," preferring to keep company with "lonely soldiers" than to sitting "in a stuffy church while an elderly priest espoused the virtues of a pure life." Too young for the USO, Ponce enjoyed going to *"las vistas,"* usually singing cowboy movies shown in the church hall after Sunday evening rosary.[64]

Priests endeavored to provide wholesome entertainment, showing films approved by the Legion of Decency. Movies in parish halls also served other purposes. The cut-rate features, like church *jamaicas,* raised money for local activities and offered a social space for parishioners. In an era of segregated theatres, church halls tendered an environment where Mexicanos and their children could enjoy inexpensive entertainment and sit wherever they pleased.[65] Even within the fishbowl of church-sponsored functions, romance could blossom. In Riverside, Frederico Buriel kept going to the movies held every Sunday night at Our Lady of Guadalupe Shrine so he could chat with the pretty ticket seller, Eusebia Vásquez. Theirs, however, was not a teenage courtship. When they married, she was thirty-seven, he forty-three.[66]

Parents could also rely on Catholic practices in the home to test the mettle of prospective suitors. When Fermín Montiel came to call on Livia León in Rillito, Arizona, her parents instructed him to join them as they knelt to recite the family rosary. In Livia's words: "It was a real education for him to be told it was rosary time."[67]

As a manifestation of familial oligarchy, chaperonage crossed denominational lines. Protestant teens, too, yearned for more freedom of movement. "I was beginning to think that the Baptist church was a little too Mexican. Too much restriction," remembered Rose Escheverria Mulligan. Indeed, she longed to join her Catholic peers who regularly attended church-sponsored dances—"I noticed they were having a good time."[68]

As mentioned earlier, popular culture offered an alternative vision to parental and church expectations complete with its own aura of legitimacy. While going out with a man alone violated Mexican community norms, such behavior seemed perfectly appropriate outside the barrio. Certainly Mexican-American women noticed the less confined lifestyles of their Anglo co-workers who did not live at home and who went out on dates unchaperoned. Some wage-earning teenagers rented apartments, at times even moving in with Anglo peers. Both English and Spanish-language media promoted a freer heterosocial environment. Radios, magazines, and movies held out images of neckers and petters, hedonistic flappers bent on a good time. From Middletown to East Los Angeles, teenagers across class and ethnicity sought to emulate the fun-seeking icons of a burgeoning consumer society.[69]

Even the Spanish-language press fanned youthful passions. On May 9, 1927, *La Opinion* ran an article entitled, "How do you kiss?" Informing readers that "el beso no es un arte sino una ciencia" [kissing is not an art but rather a science], this short piece outlined the three components of a kiss: quality, quantity, and topography. The modern kiss, furthermore, should last three minutes.[70] Though certainly shocking older Mexicanos, such titillating fare catered to a youth market. *La Opinion,* in many respects, reflected the coalescence of Mexican and American cultures. While promoting pride in Latino theater and music, its society pages also celebrated the icons of Americanization and mass consumption.

Mexican-American women were not caught between two worlds. They navigated across multiple terrains at home, at work, and at play. They engaged in cultural coalescence. The Mexican-American generation selected, retained, borrowed, and created their own cultural forms. Or as one woman informed anthropologist Ruth Tuck, "Fusion is what we want—the best of both ways."[71] These children of immigrants may have been captivated by consumerism, but few would attain its promises of affluence. Race and gender prejudice as well as socioeconomic segmentation constrained the possibilities of choice.

The adult lives of the seventeen narrators profiled in this chapter give a sense of these boundaries. Most continued in the labor force, combining wage work with household responsibilities. Their occupations varied from assembling airplanes at McDonnell-Douglas to selling clothes at K-Mart.[72] Seven of the seventeen married Euro-American men, yet, their economic status did not differ substantially from those who chose Mexican partners.[73] With varying degrees of financial security, the majority of the narrators are working-class retirees whose lives do not exemplify rags to riches mobility, but rather upward movement within the working class. Although painfully aware of prejudice and discrimination, many people of their generation placed faith in themselves and faith in the system. In 1959, Margaret Clark asserted that the second-generation residents of Sal si Puedes [a northern California barrio] "dream and work toward the day when Mexican Americans will become fully integrated into American society at large."[74] Perhaps, as part of that faith, they rebelled against chaperonage.

Indeed, what seems most striking is that the struggle over chaperonage occurred against a background of persistent discrimination. During the early 1930s, Mexicans were routinely rounded up and deported and even when deportations diminished, segregation remained. Historian Albert Camarillo has demonstrated that in Los Angeles restrictive real estate covenants and segregated schools increased dramatically between 1920 and 1950. The proportion of Los Angeles area municipalities with covenants prohibiting Mexicans and other people of color from purchasing residences in certain neighborhoods climbed from 20 percent in 1920 to 80 percent in 1946. Many restaurants, theaters, and public swimming pools discriminated against their Spanish-surnamed clientele. In southern California, for example, Mexicans could swim at the public plunges only one day out of the week (just before they drained the pool).[75] Small-town merchants frequently refused to admit Spanish-speaking people into their places of business. "White Trade Only" signs served as bitter reminders of their second-class citizenship.[76]

Individual acts of discrimination could also blunt youthful aspirations. Erminia Ruiz recalled that from the ages of thirteen to fifteen, she worked full-time to support her sisters and widowed mother as a doughnut maker. "They could get me for lower wages." When health officials would stop in to check the premises, the underage employee would hide in the flour bins. At the age of sixteen, she became the proud recipient of a Social Security card and was thrilled to become the first Mexican hired by a downtown Denver cafeteria. Her delight as a "salad girl" proved short-lived. A co-worker reported that $200 had been stolen from her purse.[77] In Erminia's words:

> Immediately they wanted to know what I did with the $200.00. I didn't know what they were talking about so they got . . . a policewoman and they took me in the restroom and undressed me. [Later they would discover that the co-worker's friend had taken the money.] I felt awful. I didn't go back to work.[78]

Though deeply humiliated, Erminia scanned the classified ads the next day and soon combined work with night classes at a storefront business college.[79]

During the course of her adolescence, Erminia Ruiz was the sole breadwinner. Even so, familial oligarchy significantly influenced her actions. For Erminia, there was no *dueña* because she was not allowed to date at all. However, she found ways to have a social life. On weekend evenings, she had to meet her dates at the drugstore or sneak out to go with girlfriends to the Rainbow Room. Though happy to be learning shorthand, Erminia dreamed of being a teacher. When an elderly patron of a cafeteria where she worked offered the financial support needed for her to complete a high school and then a college education, Erminia's mother refused. "All she said was—'I need her to work.'"[80]

Mexican-American adolescents felt the lure of Hollywood and the threat of deportation, the barbs of discrimination, and the reins of constant supervision. In dealing with all the contradictions in their lives, many young women focused their attention on chaperonage, an area where they could make decisions. The inner conflicts expressed in the oral histories reveal that such decisions were not made impetuously. Hard as it was for young heterosexual women to carve out their own sexual boundaries, imagine the greater difficulty for lesbians coming of age in the Southwest barrios.

Although only one facet in the realm of sexual politics, chaperonage was a significant issue for teenagers coming of age during the interwar period. The oral interviews on which I relied represented the experiences of those who were neither deported nor repatriated. Future studies of chaperonage should examine the experiences of those adolescents whose families returned to Mexico. This physical uprooting no doubt further complicated the ways in which they negotiated across shifting cultural terrains. For example, were young women expected to tolerate even more stringent supervision in Mexico? Did they, in essence, return to the ways of their mothers or did they adopt distinctive patterns of behavior depending on which side of the political border they found themselves?

Future research should also look more closely at the worlds of Mexican mothers. Were they, in fact, the tradition-tied Mexicanas who longed for a golden age of filial obedience or did they too hear the siren song of consumption? Recent studies on European immigrants by Susan Glenn and Andrew Heinze suggest that U.S. popular culture made inroads among mothers as well as daughters. Based on my previous work, I would argue that employment outside the home would accelerate the process of acculturation among immigrant women.[81] Interestingly, several Mexican actors attributed the decline of Spanish language theater in the Southwest during the Great Depression to financial hard times, repatriation drives, *and* the Americanized tastes of Mexican audiences.[82]

Coming of age during the interwar period, young women sought to reconcile parental expectations with the excitement of experimentation. Popular culture affirmed women's desire for greater autonomy and, in hearing its messages, they acted. Chaperones had to go.

In studying the interwar generation, a pattern emerges regarding the presence of *la dueña*. Although still practiced in some areas, chaperonage appeared less frequently after World War II. By the 1950s, chaperonage had become more of a generational marker. Typically only the daughters of recent immigrants had to contend with constant supervision. Mexican Americans relegated chaperonage to their own past, a custom that, as parents, they chose not to inflict on their children. Family honor also became less intertwined with female virginity; but the preservation of one's "reputation" was still a major concern.[83] In the poem "Pueblo, 1950," Bernice Zamora captures the consequences of a kiss:

> I remember you, Fred Montoya
> You were the first *vato* to ever kiss me
> I was twelve years old.
> My mother said shame on you,

> my teacher said shame on you, and
> I said shame on me, and nobody
> said a word to you.[84]

Some Mexican-American women found themselves invoking the threats of their mothers. María de las Nieves Moya de Ruiz warned her daughter Erminia that if she ever got into trouble and disgraced the family, she would be sent packing to a Florence Crittenden Home for Unwed Mothers. During the 1960s and 1970s, Erminia repeated the same words to her teenage daughters. Chaperonage may have been discarded, but familial oligarchy remained.[85]

In challenging chaperonage, Mexican-American teenagers did not attack the foundation of familial oligarchy—only its more obvious manifestation. (It would take later generations of Chicana feminists to take on this task.) Chaperonage, however, could no longer be used as a method of social control, an instrument for harnessing women's personal autonomy and sexuality. Through open resistance and clever evasion, daring young women broke free from its constraints. Their actions represent a significant step in the sexual liberation of Mexican-American women.

4

With Pickets, Baskets, and Ballots

◆※◆

> A few nights ago I spoke to 1,500 women—women who work
> picking walnuts out of shells. It was one of the most amazing
> meetings I've ever attended. . . . The employers recently took
> their hammers away from them—they were making "too much
> money." For the last two months . . . they have been cracking
> walnuts with their fists. Hundreds of them held up their fists to
> prove it.[1]

CAREY McWilliams wrote these lines in a 1937 letter to his friend
Louis Adamic. Describing these Eastern European and Mexican
workers, he observed, "And such extraordinary faces—particularly
the old women. Some of the girls had been too frequently to the
beauty shop, and were too gotten up—rather amusingly dressy."[2] I
would argue that dressing up for a union meeting could be interpreted as an affirmation of individual integrity. They had not sur-
rendered their self-esteem as evidenced by their collective action
and personal appearance. These women worked in sweatshops.
Shells littered the shop floor, causing them to slip and fall, and the
toilets were "filthy." Less than a year later, the walnut workers
joined the United Cannery, Agricultural, Packing, and Allied Workers of America (UCAPAWA-CIO). Dorothy Ray Healey, UCAPAWA
vice-president and Los Angeles organizer, recalled that the "catalyst" for labor representation seemed unrelated to either hammers
or wages, but work benches. "The work benches they stood in front
of . . . were ragged and jagged and tore their stockings. Every day

they had to get another pair of stockings and that just infuriated them."[3] Conforming to popular fashions and fads cannot be construed as a lack of ethnic or political consciousness. In this instance, silk stockings were accessories to unionization.

Spanish-speaking women as family members, as workers, and as civic-minded individuals have strived to improve the quality of life at work, home, and neighborhood. This chapter does not provide a laundry list of every strike, *mutualista*, or political organization in which Mexican women have exercised leadership. Instead it offers a range and a sample of such activist paths, drawing out, in particular, the integration of women's private and public worlds, worlds in which family and community have been (at least) metaphorically intertwined. As farm hands, cannery workers, miners' wives, *mutualista* members, club women, civil rights advocates, and politicians, Mexican women have taken direct action for themselves and others. This claiming of public space has occurred at several levels and with differing trajectories across time and region. Some, like the walnut workers, placed their faith in a union local; some joined *mutualistas*; and others, such as Alicia Dickerson Montemayor of LULAC, turned to the ballot box.[4]

Reflecting the heterogeneity within Mexican and Mexican-American cultures as well as individual predilections and social locations, political agendas have varied. For example, in 1939, Luisa Moreno, Josefina Fierro de Bright, and Eduardo Quevedo were the principal leaders of El Congreso de Pueblos de Hablan Española (the Spanish-speaking Peoples Congress), the first national Latino civil rights assembly. Two years later, on the more conservative side of the spectrum, New Mexico politician and community leader Concha Ortiz y Pino railed against welfare administrators whom she viewed as spendthrift social workers.[5] At times men and women holding diverse political views joined together in common cause, particularly in areas as voter registration.

This chapter sketches out the contours of Mexican women's very public roles, roles they often assumed in pursuit of social justice and in partnership with men. As historian Cynthia Orozco has so eloquently stated:

> For over a century Chicanas have belonged to voluntary associations. We too have a history of voluntarist politics but our scholarly discourse on the subject is not about "volunteerism" or "voluntarist politics" but rather in the discourse of "community organizations," "activism," and the politics of "resistance."[6]

I further concur with Orozco's observations on the significance of mixed gender associations among Mexicans in the United States.[7] These legacies of resistance encompassed both compadres and co-madres.

Mexican women have a rich history of union and political activism. As far back as 1903, they took matters in their own hands in a labor dispute. Rallying to support 700 striking track workers of the Pacific Electric Railway Company in Los Angeles, Mexicanas caught the attention of a local reporter. In his words:

> The women had come from various parts of Sonoratown and all of them were relatives or sympathizers with the Mexicans who had gone on strike. . . . There were more than thirty . . . Amazons . . . [they] approached the workers [scabs] and began seizing the shovels, picks, and tamping irons.[8]

Referring to the English working class, legendary scholar E. P. Thompson wrote that "there was a consciousness of the identity of interests . . . which was embodied in many institutional forms, and which was expressed on an unprecedented scale in . . . general unionism."[9] During the 1930s, Mexican workers had reached a similar stage. In 1933 alone, thirty-seven major agricultural strikes occurred in California—twenty-four led by the Cannery and Agricultural Workers Industrial Union (CAWIU). The leadership of this union identified with the Communist Party—from principal officers Pat Chambers and Caroline Decker to grass-roots organizers Elizabeth Nicholas and sixteen-year-old Dorothy Ray (later Healey).[10] Daring to organize some of the most disenfranchised people in the United States, these activists enjoyed an enviable success rate, winning partial wage increases in twenty-one of twenty-four disputes. The San Joaquin Valley Cotton Strike was the union's most ambitious attempt. Between 12,000 and 20,000 workers, spanning a 120-mile area, engaged in a bitter dispute over wages. They wanted their pay increased from sixty cents to one dollar per hundred pounds of cotton picked. While the union provided the umbrella leadership, Mexican families composed 95 percent of the rank and file.[11] It was not as if the union were calling all the shots and the workers followed. As Pat Chambers told Devra Weber, "Although the directives in some superficial way could come from the outside, the actual organization had to come from the workers themselves."[12]

Violence marred the labor dispute. Farmers killed three people, including one Mexicana. Law enforcement officials, as well as the local press, threatened strikers with deportation. Along with grower and police harassment and outright terror, a denial of federal relief proved an effective strike-breaking method. One Tulare County supervisor allegedly stated, "Those d—— Mexicans can lay out on the street and die for all I care." Indeed, at least four children died of malnutrition during the struggle. In government hearings held several years later, a Kern County deputy sheriff showed no remorse for police actions. In his words, "We protect our farmers here. . . . They are our best people. . . . But the Mexicans are trash. . . . We herd them like pigs." After twenty-four days, union members, possessing few options, reluctantly accepted a fifteen-cent wage increase.[13]

Women did not stand on the sidelines. They distributed food, formed picket lines, taunted scabs, and, when attacked by police, fought back. Devra Weber described them as "older women with long hair who wore the rebozos of rural Mexico, young women who had adopted flapper styles, and young girls barely in their teens."[14] As they did in labor camps and colonias, women's networks offered physical and emotional support. As channels for political and labor activism, they also fused private life and public space in pursuit of social justice.

The San Joaquin Valley Cotton Strike represents the "David versus Goliath" image of farm workers organizing—Mexican *campesinos* versus Euro-American agribusiness. The El Monte Berry Strike of 1933 presents a different, more complicated scenario. On June 1, 1933, 1,500 berry pickers went out on strike in the fields surrounding El Monte, California. Organized initially by the CAWIU, the workers (the majority of them Mexicano) demanded a pay increase because some earned as little as nine cents per hour. Jesusita Torres, who did this work, remembered earning less than one penny per basket of berries. The employers were not Euro-American farmers, but Japanese leaseholders.[15] Consequently, the relationship between grower and worker did not appear as cut and dried or as distant. Within the cultural landscape of El Monte, both groups were minorities—numerically and socially.

El Monte was a small town with a population under 10,000 residents (75 percent Euro-American; 20 percent Mexican; and 5 percent Japanese) and one with clearly marked racial divisions in housing, schools, and public facilities. Mexicans and Japanese were segregated from Anglo El Monte. The children of Japanese farmers and Mexican farm workers attended the same segregated school, Lexington Elementary, and in the town's premier movie palace, they were relegated to the same side of the aisle, away from Anglo patrons. This sharing of social space in the classroom or the cinema led to an environment in which grower-*campesino* relations were familiar, but not friendly. According to Señora Torres, "They [the Japanese farmers] would work in the field, but you knew they were the *boss*."[16]

Historian Rudy Acuña argues that the Japanese leaseholders were "caught" between rising rents levied by Euro-American landowners and worker protests for higher wages. I would add that Mexican strikers were also caught between two rival unions—the class-based CAWIU and the nationalist CUCOM (La Confederación de Uniones de Campesinos y Obreros del Estado de California), a union supported by the Mexican Consul. Once the rank-and-file leadership switched to CUCOM, over $5,000 in strike support funds flowed from the pockets of Mexico's politicians, including hefty donations from the nation's current and former presidents. Fearing adverse publicity for his compatriots and their children, the Japanese Consul stepped in to try to broker a settlement. Such international intervention in a U.S. labor dispute seemed unprecedented. Furthermore, the chair of the strike committee, Armando Flores, sought federal help as he appealed to President Roosevelt for arbitration under the new National Industrial Recovery Act (NIRA).[17]

Yet the success or failure of the strike would depend, in part, on the strength of community networks. The residents of El Monte's Hick's Camp barrio rallied around the strikers. Sadie Castro used her culinary skills to feed hungry families. "She used to cook rice and beans and take it to the people who were striking," recalled Patty Holguin. Barrio merchants also donated food and other needed supplies. Strikers who became scabs, moreover, found themselves ostracized by their neighbors.[18]

Farmers also relied on their own networks. From all over the Los Angeles area, family and friends pitched in to harvest the ripe and highly perishable berries. "Japanese school children were released from school for a few days to work in the fields." Growers also sought to recruit the public as scab labor. Newspaper ads and radio spots offered El Monte berries at the pick-your-own price of one penny per box. With thoughts of bargain berries, "hundreds of men, women and children" took a trip to the fields.[19]

On July 6, a settlement was reached: "$1.50 for a nine-hour day or 20 cents an hour where the employment was not steady." This agreement resulted largely from the concerted efforts of the state and federal Departments of Labor, the Mexican and Japanese Consuls, and the Los Angeles Chamber of Commerce. Ethnic-based mutual aid played a significant role in that both sides relied on their racial/ethnic communities to help them weather the strike. In 1946, sociologist Charles Spaulding contended that the strike had widened the gulf between Mexican workers and Japanese growers while shrinking the "social distance" between Japanese and Euro-American residents of El Monte. Coalitions based on class interests or affinities perhaps precluded the development of racial/ethnic coalitions to fight segregation. After Pearl Harbor, El Monte growers would find themselves judged not by class, but by race. Surprisingly, Spaulding ignores the internment of Japanese Americans during World War II and its impact on both area growers and workers. As Jesusita Torres related, "After they were taken to the concentration camps, the fields were not good."[20]

When one thinks of California agriculture during the Depression, John Steinbeck's *The Grapes of Wrath* generally comes to mind. Steinbeck did not exaggerate. White and African American Dust Bowl migrants alongside Mexicanos and Filipinos endured great hardships. Local relief agencies in the San Joaquin Valley turned their back on the plight of farm workers, and growers continued to lower piece rates confident in the knowledge that migrants would take any job under any conditions.[21] Beginning in 1937, UCAPAWA-CIO offered hope. Four decades later, I asked Dorothy Ray Healey to recall her most rewarding experience as a labor organizer. Her answer:

> To watch the disappearance or at least the diminishing of bigotry . . . watching all those Okies and Arkies and that . . . bigotry and small-mindedness—all their lives they'd been on a little farm in Oklahoma; probably they had never seen a Black or a Mexicano. And you'd watch in the process of a strike how those white workers soon saw that those white cops were their enemies and that the Black and Chicano workers were their brothers.[22]

Borne out of material conditions, class consciousness provided the leavening for labor activism. Mexican men and women proved enthusiastic union members. Some acted out of socialist convictions nurtured during the Mexican Revolution; others viewed the union solely as a vehicle for decent wages and conditions. Whether envisioning class struggle or the American Dream, migrant families responded to union overtures to such a degree that the *San Francisco*

News declared that UCAPAWA was the fastest growing agricultural union in California history primarily because of to its popularity among Mexicanos.[23] UCAPAWA capitalized on women's networks, with organizers holding house meetings to encourage their participation.

The Associated Farmers (AF), a vigilante anti-union brigade of growers and their supporters, also detected this militancy among Spanish-speaking workers. As one grower malevolently wrote, "We do not propose to sit idly by and see the fruits of our labors destroyed by a bunch of Indian ignoramusses from the jungles."[24] Hired by both the AF and Fresno County law enforcement, a private investigator meticulously recorded in his mileage reports the names of various Mexican cafes he considered "Communist and CIO" hotbeds.[25]

Given the union's growing membership and partial victories, the Associated Farmers swung into action—often relying on violence to achieve their goals as evident in the Madera Cotton Strike. On October 12, 1939, approximately 1,000 white, Mexican, and African-American pickers walked out of the fields around Madera County; it was in one grower's estimation about 90 percent of the workforce. Thirty Fresno ministers publicly encouraged area farmers to raise their pay scales so families would not have to depend on the wages of women and children. Family members also shared picket duty and when police arrested 143 strikers for violating the local anti-picketing ordinance, they were incarcerated together.[26] Led by the AF, local growers sprinkled tacks onto roadways and beat picketers with clubs and chains. When 200 vigilantes attacked UCAPAWA families attending a peaceful rally in the park, law enforcement officials stood back to watch and eventually entered the fracas by "by shooting tear gas into the park to rout the strikers." Nineteen union members required medical attention.[27] An outside observer would write, "Men, women, and little children with nowhere to sleep, nothing to eat, are hunted, shot, and beaten because they asked for a wage they could live on."[28]

Unlike previous attempts to organize field workers, UCAPAWA enjoyed the support of several outside groups such as the John Steinbeck Committee to Aid Agricultural Organization, the Simon J. Lubin Society, and the Spanish-speaking Peoples Congress. Hollywood celebrities threw Christmas parties for farm workers and city folk brought foodstuffs and other goods they had collected on their behalf. Although the Madera strikers would realize their minimum wage demands, UCAPAWA would soon retreat from the fields. The tactics of the Associated Farmers, local hostility to the union, the failure of the New Deal to extend collective bargaining rights to agricultural workers, and the drain on union coffers necessitated a new strategy. UCAPAWA activists would turn their attention to cannery and packinghouse workers on both national and regional levels with the idea of building stable food-processing locals from which to base a renewed organizing campaign among migrant field hands.[29] The union did not forget the militancy of Mexicanos, especially of women, and from San Antonio pecan-shelling plants to Los Angeles peach canneries, UCAPAWA would fuse class consciousness with gendered meanings—capitalizing on women's networks for unionization as well as encouraging the leadership skills of rank-and-file women.

Between 1933 and 1938, Mexican workers organized a pecan shellers' union, El Nogal, in San Antonio, Texas. Men, women, and children were paid pitifully low wages—less than $2.00 a week in 1934. Management conceded that wages were low, but contended that the shellers could eat all the pecans they desired and chat with friends while they worked. Some employers explained that if they raised the pay scale, Tejanos would "just spend" the extra money on "tequila and worthless trinkets in the dime stores."[30] One employer, Julius Seligman, defended the pay scale of five cents per day even though his company had garnered over $500,000 in profits.[31]

As a twenty-three-year-old member of the Workers' Alliance and secretary of the Texas Communist Party, Emma Tenayuca emerged as the fiery local leader. Although not a pecan sheller, Tenayuca, a San Antonio native, was elected to head the strike committee. During the six-week labor dispute from 6,000 to 10,000 strikers faced tear gas and billy clubs "on at least six occasions." Emma Tenayuca courageously organized demonstrations and she along with over 1,000 pecan shellers were jailed.[32] Known as "La Pasionaria," Tenayuca, in an interview with historian Zaragosa Vargas, reflected on her activism as follows: "I was pretty defiant. [I fought] against poverty, actually starvation, high infant death rates, disease and hunger and misery. I would do the same thing again."[33]

UCAPAWA president Donald Henderson intervened in the strike by assigning Luisa Moreno, a thirty-two-year-old veteran labor activist from the East Coast, to San Antonio to help solidify the local and move from street demonstrations to a functioning trade union. As the union's official representative, Moreno organized the strikers into a united, disciplined force that employers could no longer ignore. Five weeks after the strike began, management agreed to arbitration. The settlement included both recognition of the UCAPAWA local and piece-rate scales that would comply with the newly established federal minimum wage of twenty-five cents an hour.[34]

Employers reacted to unionization by installing new equipment that performed most of the manual tasks. From 1938 to 1950, employment in the pecan-shelling plants fell from 10,000 to 350 workers. The wages of the operatives, however, remained reasonably stable. According to historian Robert Landolt, a single pecan sheller during the 1940s earned "as much or more than the labor of an entire family" working in the plants prior to unionization. Side benefits included greater opportunities for Mexican children to attend school rather than labor in a pecan sweatshop to help feed their families.[35]

Tejana pecan shellers held positions of leadership in their union from the onset of the strike through World War II. Three of the workers at the bargaining table in 1938 were women and in 1942 Tejanas held all the top spots in UCAPAWA Local No. 172, The Pecan Shellers Union of San Antonio. Lydia Dominguez served as president, Margarita Rendón as vice-president, and Maizie Tamez as secretary-treasurer.[36] UCAPAWA's most successful union drives among Mexican women workers, however, would not be in Texas, but in the canneries and packinghouses of sunny southern California. Women's networks in southern California canneries at times crossed ethnic boundaries and organizers quickly recognized their potential as conduits for unionization.

During the 1930s, the canning labor force included young daughters, newly married women, middle-aged wives, and widows. Occasionally three generations worked at a particular cannery— daughter, mother, and grandmother. Entering the job market as members of a family wage economy, they pooled their resources to put food on the table. "My father was a busboy," Carmen Bernal Escobar recalled, "and to keep the family going . . . in order to bring in a little more money . . . my mother, my grandmother, my mother's brother, my sister and I all worked together at Cal San."[37]

One of the largest canneries in Los Angeles, the California Sanitary Canning Company (Cal San) employed primarily Mexican and Russian Jewish women. They were clustered into specific departments—washing, grading, cutting, canning, and packing—and were paid according to their production level. Women jockeyed for position near the chutes or gates where the produce was plentiful. "Those at the end of the line hardly made nothing." Standing in the same spots week after week, month after month, women workers often developed friendships crossing family and ethnic lines. Their day-to-day problems (slippery floors, peach fuzz, production speed-ups, arbitrary supervisors, and sexual harassment) cemented feelings of solidarity. Cannery workers even employed a special jargon when conversing among themselves, often referring to an event in terms of when specific fruits or vegetables arrived for processing at the plant. For instance, the phrase, "We met in spinach, fell in love in peaches, and married in tomatoes" indicates that the couple met in March, fell in love in August, and married in October.[38]

In addition, second-generation daughters shared celebrity *chisme, True Story,* and Pond's cold cream. They had common interests rooted not only in their work but in their positionality as second-generation women coming of age during the Depression. Not every Mexican-American teenager acquired a Euro-American work buddy, but enough women transcended barriers of mutual distrust and wage disparities so that, at certain junctures, the parallel networks met and collective strategies, such as unionization, could be created and channeled across ethnic boundaries.[39]

In July 1939, Dorothy Ray Healey, now a vice-president for UCAPAWA, began to distribute union leaflets outside the gates of Cal San. Meeting were held in workers' homes so that entire families could listen and membership cards traveled from one kin or peer network to the next. Within three weeks, 400 (out of 430) Cal San employees had joined UCAPAWA. The cannery owners, the Shapiro brothers, refused to recognize the union and a strike was called. At the height of the peach season (August 31, 1939), 400 workers walked off their jobs, the next day sixteen more joined them. The workers staged picket lines around the plant, at local grocery stores that carried Cal San products, and eventually their children would picket the front lawns of the Shapiros' homes. Within days of the child pickets who carried such signs as "I'm underfed because my Mama is underpaid," the Shapiros agreed to meet with the negotiating team and a settlement was quickly reached.[40]

Wages and conditions improved at Cal San as workers nurtured their local and jealously guarded their closed shop contract. In 1941, Luisa Moreno arrived to organize other canneries in southern California. She enlisted the aid of union members at California Walnut and Cal San in union drives at several Los Angeles area food-processing firms. The result would be Local 3, the second largest UCAPAWA affiliate in the nation. Moreno encouraged cross-plant alliances and women's leadership. In 1943, for example, Mexican women filled eight of the fifteen elected positions of the local. The union members proved able negotiators during annual contract renewals. The local also offered benefits that few industrial unions could match—free legal advice and a hospitalization plan.[41]

In southern California, UCAPAWA provided women cannery workers with the crucial "social space"[42] necessary to assert their independence and display their talents. They were not rote employees, numbed by repetition, but women with dreams, goals, tenacity, and intellect. Unionization became an opportunity to demonstrate their shrewdness and dedication to a common cause. Mexicanas not only followed the organizers' leads, but also developed strategies of their own. A fierce loyalty developed as the result of rank-and-file participation and leadership. Forty decades after the strike, Carmen Bernal Escobar declared, "UCAPAWA was the greatest thing that ever happened to the workers at Cal San. It changed everything and everybody."[43]

The World War II era ushered in a set of new options for Mexican women wage earners. Not only did cannery workers negotiate higher wages and benefits, but many obtained more lucrative employment in defense plants. As "Rosie the Riveters," they gained self-confidence and the requisite earning power to improve their standard of living. Carmen Chávez and her sisters, Albuquerque Hispanas, entered the workforce for the first time during the war. "[T]hrough earning our own wages, we had a taste of independence we hadn't known before the war," Chávez wrote. "The women of my neighborhood had changed as much as the men who went to war. We developed a feeling of self-confidence and a sense of worth." As a single parent, Alicia Mendeola Shelit purchased her first home as the result of her employment with Douglas Aircraft in southern California. Her mother cared for her children as well as cooked and cleaned. As Alicia recalled, "All I knew was just bring the money in to feed my kids, like a man."[44]

Though women could now earn more money, old problems remained. Julia Luna Mount left her cannery job at Cal San both for better pay as a C-47 framer at Douglas Aircraft and for a chance to organize with the United Auto Workers. However, she found the other Rosies cool to the idea of a union local and, understandably, felt unnerved by the sexual advances of several men at Douglas. Finally, an incident occurred that forced her out of the plant.

It was real dark and this man got very fresh . . . [he] started to chase me. I didn't want him to hug me. And he got kind of rough and I got kind of panicked.[45]

Her co-worker literally chased her around the foggy parking lot. Dropping her lunch pail, she took off at a dead run, not stopping until she had reached the Personnel Department where she quit on the spot.[46]

While Julia Luna Mount was physically threatened, other Mexican-American Rosies would have their reputations paraded through the mud in the popular press. These young women, known collectively as *"las pachucas,"* rebelled against Mexican and American norms. Patricia Adler commented that *pachucas* "scandalized the adults of the Anglo and Mexican communities alike with their short, tight skirts, sheer blouses, and built-up hairdos." Conversely, Chicana artist Carmen Lomas Garza in "Pachuca With a Razor Blade" created a portrait of teenagers getting dressed for a dance with one young woman concealing a razor blade in a barrette, no doubt, to be hip and protected.[47] Enjoying the company of men who wore zoot suits, *las pachucas* were vilified as incorrigible delinquents by the Los Angeles press. One daily, according to Carey McWilliams, ran a story that characterized *pachucas* as pot-smoking prostitutes with VD. Similarly, *La Opinion* referred to them as *"las malinches"*—traitors to established Mexican codes of chaperoned feminine conduct.[48] Responding to such lurid depictions, a group of East Los Angeles teens wrote a letter published in *The Eastside Sun* in which they affirmed their virginity and patriotism. A portion follows:

> The girls in this meeting room consist of young girls who graduated from high school as honor students, of girls who are now working in defense plants because we want to help win the war, and of girls who have brothers, cousins, relatives, and sweethearts in all branches of the American armed forces. We have not been able to have our side of the story told.[49]

In 1942, the director of the Civilian Service Corp, Herman Stark, tried to set the record straight. He declared that juvenile delinquency in Los Angeles has been "grossly exaggerated" in that "only 1.6 percent of boys and .6 percent of girls" required "the attention of juvenile authorities." The English-language press placed the blame for the Zoot Suit Riots of 1943 not on the Euro-American servicemen who indiscriminately attacked Mexican youth on Los Angeles street corners, but on the zoot suiters themselves.[50] Statements by Herman Stark and the East Los Angeles teenagers stand in stark contrast to popular conceptions. The letter printed in *The Eastside Sun* reflects the positionality of teenagers who as wage workers and citizens sought to carve out their own social space, not in terms of exercising union leadership, but by defining a youth culture.[51]

There were reasons for hope and fear. During the 1940s, Mexican women began to break into lower white-collar occupations. In 1930, only 10 percent of Mexican women workers in the Southwest held clerical or sales positions, but by 1950 this figure had risen to 23.9 percent. With the expansion of clerical jobs during the war, second-generation daughters could finally apply the office skills they had acquired in high school or at storefront business colleges.[52] The Zoot Suit Riots of June 1943 and Japanese-American internment, however, served as reminders of their own status as second-tier citizens. Civil rights leader Josefina Fierro de Bright witnessed the riots firsthand. When she pleaded with a police officer to intervene, he told her, "Keep your mouth shut or they're gonna beat the sh— out of you, too." Josefina further recalled that *pachucos* were not the only targets of soldiers and sailors as men in uniform roamed the aisles of local Mexican theaters assaulting men, women, and children. Diana Bernal lived in San Antonio, Texas, and Theresa Negrete in Scottsbluff, Nebraska, but both women feared that Mexicans would be the next group placed in concentration camps. As Bernal stated, "Are they going to do the same thing to us because we are Mexican?"[53]

The postwar era did little to promote a more tolerant environment, although cracks in segregation began to surface. In 1946, federal judge Paul McCormick ruled in *Mendez v. Westminister* that the segregation of Mexican schoolchildren was unconstitutional; his interpretation of the Fourteenth Amendment would later serve as a precedent for *Brown v. Board of Education.* However, cannery workers in southern California lost their union as UCAPAWA/FTA collapsed under the weight of Red scare witch hunts exemplified by the California Senate UnAmerican Activities Committee and Teamster smear campaigns. In 1950, Luisa Moreno left for her native Guatemala under terms listed as "voluntary departure under warrant of deportation" on the grounds that she had once been a member of the Communist Party.[54]

The movie *Salt of the Earth* encapsulated Cold War politics and

Mexican-American labor activism. Produced, directed, and written by filmmakers blacklisted from Hollywood, the film recorded the real-life struggles of Mexican miners in Silver City, New Mexico, who walked off their jobs over wage and safety issues. The strike began in October 1950 and lasted until January 1952. Organized by Clint and Virginia Jencks, who represented the International Union of Mine, Mill, and Smelter Workers, the miners faced police harassment and were prohibited from picketing. Their wives took their places on the line and men assumed child care and household responsibilities, learning to cope without hot water or indoor plumbing. Although assaulted and arrested, women refused to be intimidated. Mariana Ramírez described their arsenal against the scabs. "We had knitting needles. We had safety pins. . . . We had chili peppers." Another perhaps disingenuously informed a local reporter, "It's like a picnic. . . . We're having fun—and we're going to stay on the picket line, too." The movie documents the changes in consciousness about women's place within mining families as a result of the temporary role reversals.[55]

At Empire Zinc, Clint and Virginia Jencks cultivated women's leadership in ways that closely resembled UCAPAWA's efforts in canneries and packinghouses. Indeed, three of the five members of the union negotiating team were not miners, but wives who held the line—Elvira Molano, Catalina Barreras, and Carmen Rivera. An arbitrated settlement was reached in favor of the mine families in one of the few labor victories during the 1950s, a successful strike that achieved both higher wages and hot water.[56]

During the filming of *Salt of the Earth,* the cast and crew ran the proverbial gauntlet. Megamillionaire Howard Hughes used his influence to bar the doors to Hollywood's film labs and editing rooms. Even before the picture was released, Republican Congressman from California Donald Jackson publicly referred to it as "deliberately designed to inflame racial hatred . . . a new weapon for Russia." The Immigration and Naturalization Service harassed the principal star, Mexican actor Rosaura Revueltas, throughout the filming and toward the end of the shoot she was arrested and held for a deportation hearing. In describing the INS prosecutor, Revueltas explained, "Since he had no evidence to present of my 'subversive' character, I can only conclude that I was 'dangerous' because I had been playing a role that gave status and dignity to the character of a Mexican-American woman." She left the United States under a warrant of deportation. In *Salt of the Earth: Story of a Film,* director Herbert Biberman offers a spine-tingling tale of courage, conviction, and Cold War repression.[57] Attending a thirty-year *Salt of the Earth* reunion, Angela Sánchez, who was both picket and movie extra, recalled two lasting legacies. "Mexican culture has traditionally been that women belong in the home. After the strike, men didn't see us as weaklings anymore. Also the two races started getting together and mixing in nightclubs."[58] The impact of the strike on race and gender relations certainly varied from family to family, yet its celluoid testament in *Salt of the Earth* remains emblematic of a long history of labor activism among Mexican women in the United States.

If holding the picket line represented one way Mexican women claimed a collective public space, distributing food baskets was another. Women participated in *mutualistas* (mutual aid societies) that materialized in southwestern barrios during the late nineteenth century. Many early organizations focused on providing life insurance, planning Mexican patriotic celebrations, and offering charitable and legal assistance. According to Bert Corona, a Chicano trade union and civil rights leader, *mutualistas* "were often the only means immigrants had to defend themselves and to acquire help."[59] At times these societies were part of larger networks either affiliated with the Mexican Consulate or an autonomous regional confederation. Founded by prominent Hispanic citizens of Tucson in 1894, Alizana Hispano-Americana not only focused on local welfare needs but formed new chapters in Phoenix, Tempe, Jerome, and other towns in Arizona territory. By 1919, the Alianza, which had cultivated a substantial working-class following, begot eighty-eight chapters across the Southwest and northern Mexico and, on the eve of World War II, over 17,000 people were dues-paying members. One historian succinctly stated that the Alianza "hoped to unite all Mexicans and Latin Americans in the United States into one 'family' under the principles of 'protection, morality, and education.'"[60]

Mutualista "families," however, differed by citizenship, class, and gender. In Santa Barbara, Club Mexicano Independencia (CMI), a working-class community group, offered its members a range of insurance and emergency benefits; however, membership was predicated on being male and a citizen of Mexico. In contrast, Club Femenino Orquidia in San Antonio was an elite women's organization that mixed volunteer service with cultural events and lectures. One evening, its members gathered for a lecture by Margarita Robles de Mendoza sponsored, in part, by the American Association of University Women; the topic was "The Current Condition of Mexican Women."[61] The typical *mutualista* created a separate "ladies auxiliary," as did both the Alianza and CMI. As Cynthia Orozco has pointed out, women's work in mutual aid activities—preparing enchilada suppers, organizing *jamaicas* and dances as well as distributing food baskets—has received only passing mention by contemporary historians. Certainly their invisible day-to-day labor made a difference.[62]

Cultural nationalism signified voluntarist politics in the barrios of the Southwest and Midwest. *Mutualistas* were frequently named in honor of Mexican heroes—Miguel Hidalgo and Benito Juárez were popular choices. In Laredo, the popular women's group Sociedad Josefa Ortiz de Dominguez commemorated a heroine of Mexican independence.[63]

Mutualistas were important meeting grounds between Mexican immigrants and Mexican Americans and between middle-class leaders and working-class members. David Gutiérrez put it this way:

> By providing a place where immigrants and citizens of Mexican descent could speak the same language, discuss common problems, and cooperatively provide themselves with needed services, mutualistas allowed immigrants to learn the ropes of living in the United States in a . . . supportive environment. In addition . . . the culturally familiar mutualistas helped to break down barriers between the two groups, improved communication, and promoted a spirit of cooperation.[64]

This was not an unconscious process. The formation of a common identity fostered by voluntarist politics can be discerned in the words of a Houston teacher Cástulo Ortiz as he spoke before a crowd attending a Cinco de Mayo celebration:

> MEXICANS: Good will should be forged like a bell. Let it be forged in a grand effort and may a new bronze nationality, like the Aztec race come to light, bringing order to chaos. Forget political and religious divisions and come together in the future of one Mexican family.[65]

The nexus of community and family resonates throughout the history of Mexicans in the United States. Furthermore, the roots of contemporary debates over self-identification and the construction of ethnic identity can be gleaned from the very names of *mutualistas* active during the early decades of the twentieth century—

Alianza Hispano-Americana [Hispanic]; Liga Protectora Latina [Latina/o]; Sociedad Mutualista Hispano-Azteca [Chicana/o].[66]

If the politics of voluntary associations negotiated class divisions, they also articulated configurations of gendered identity. Reflecting a belief in the nurturing nature of women, Mexican Consuls across the U.S. organized chapters of La Cruz Azul (the Blue Cross), an all-women's *mutualista* devoted to improving health care, education, and social services to barrio residents. Similar to both the General Federation of Women's Clubs and the American Red Cross, La Cruz Azul could be found in such disparate locations as Los Angeles, Houston, Denver, and Detroit. Members raised funds for poor children, provided emergency relief, and established medical clinics and libraries.[67] In San Antonio, a coalition of working-class and affluent *mutualistas,* including La Cruz Azul, joined forces to build a neighborhood health clinic. Commenting on women's contributions, the local Spanish-language newspaper *La Prensa* believed that their actions could be attributed to "the latent feelings of motherhood carried by all women."[68] If La Cruz Azul was metaphorically garbed in the mantle of motherhood, the reality was one of limited resources. As historian Louise Año Nuevo Kerr has pointed out, needs always outdistanced funds and in Chicago, the chapter folded.[69]

Women, through *mutualistas,* sought to help their neighbors; they worked within their communities in a public way, although their labor generally remained invisible outside the barrio. Their activism became more evident as they joined new types of local organizations, associations that viewed the barrio not as an island of Mexico, but as a neglected American community. Groups, such as el Comité de Vecinos de Lemon Grove (the Lemon Grove Neighbors Committee), engaged in direct political action.

Before 1931 Mexican and Euro-American children in Lemon Grove, California, a small town outside San Diego, attended the same school. In January 1931, the local school board built a separate facility for Mexican pupils across the tracks in the barrio. The "new" two-room school resembled a barn hastily furnished with secondhand equipment, supplies, and books. Forming el Comité de Vecinos de Lemon Grove, local parents voted to boycott the school and seek legal redress. Except for one household, every family kept the children home. With the assistance of the Mexican Consul Enrique Ferreira, the Comité hired attorneys and filed suit. Board members justified their actions on the grounds that a separate facility was necessary to meet the needs of non-English-speaking children. To counter this argument, students "took the stand to prove their knowledge of English."[70] In *Alvarez v. Lemon Grove School District,* Judge Claude Chambers ordered "the immediate reinstatement" of Mexican children to their old school. During a reign of deportations and repatriations, Mexican immigrants had mustered the courage to protest segregation in education and they had won. Comadres and compadres banded together for grass-roots political action. These immigrants parents, moreover, had sought the assistance of the Mexican Consul in their effort to provide equal opportunities for their U.S. born children.[71]

Mutualistas represented institutionalized forms of *compadrazgo* and *commadrazgo* with men and women typically working together within parallel, if separate organizations. In Tucson, the Alianza Hispano-Americana raised funds to establish an orphanage and a nursing home, concrete manifestations of the *mutualista* as fictive kin.[72] For women, *mutualistas* represented the spaces between family and community where volunteer work was accepted and respected.[73] El Comité de Vecinos de Lemon Grove served as a bridge between mutual assistance and civil rights. During the 1930s, grass-roots organizing among men and women took several forms, with issues of class and generation defining diverse agendas.

Similar jobs, neighborhoods, and cultural traditions did not always bridge the distance between immigrants and citizens. The inclusion or exclusion of immigrant issues has served as the flash point for civil rights activists in Mexican-American communities. As David Gutiérrez explained in his pathbreaking political history, *Walls and Mirrors,* "Forced virtually every day to deal with the both the positive and negative effects of Mexican immigration while struggling to win basic civil rights for their constituents, such activists have had to come to some fundamental decisions about just who their constituents are."[74] He continues,

> Whether they choose to work only on behalf of American citizens or decided that their efforts should focus on what they considered to be a broader cultural community, including immigrants, such decisions speak volumes about Mexican American activists' political and cultural orientations in a society that has continually been transformed by constant immigration.[75]

The League of United Latin American Citizens (LULAC) chose political distance from Mexican immigrants. Founded by middle-class Tejanos in 1929, LULAC struck a chord among Mexican Americans and by 1939 chapters could be found throughout the

Southwest with a membership estimated at 2,000. Envisioning themselves as patriotic "white" Americans pursuing their rights, LULACers restricted membership to English-speaking U.S. citizens. Taking a page from the early NAACP, LULAC stressed the leadership of an "educated elite" who would lift their less fortunate neighbors by their bootstraps.[76] Mario García reiterates how members considered LULAC "an authentic American organization and that each letter in its name expressed patriotism: L stood for love of country; U for Unity as American citizens; L for loyalty to country: A for advancement; and C for citizenship." Taking a somewhat different spin, David Gutiérrez argues that "LULAC members consistently went to great lengths to explain to anyone who would listen that Americans of Mexican descent were different from (and by implication, somehow better than) Mexicans from the other side."[77] Yet both historians would agree on LULAC's concrete political legacies. According to Gutiérrez:

> From 1929 through World War II LULAC organized successful voter registration and poll tax-drives, actively supported candidates sympathetic to Mexican Americans, and aggressively attacked discriminatory laws and practices throughout Texas and the Southwest. More important . . . LULAC also achieved a number of notable legal victories in the area of public education.[78]

Indeed, landmark desegregation cases, such as *Mendez v. Westminister,* owed much to LULAC support. By 1940, LULAC had become the most visible regional civil rights organization.[79]

In their quest for respectability and political clout, LULAC members distanced themselves from working-class, *mestizo* identities. During the San Antonio Pecan Shellers Strike of 1938, LULACers sided with management and condemned the strikers. They also fought for the right to be considered "white." In his study of LULAC, political scientist Benjamin Márquez mentions a revealing incident regarding this type of racial positioning. In Corpus Christi, Texas, the city directory divided local residents into "'American' (Am), 'Mexicans' (M), 'English-speaking Mexicans' (EM), and 'Coloreds' (C)." LULACers protested their classification, incensed that they had been separated from "other members of the white race."[80] David Gutiérrez further notes that LULAC favored the restriction of Mexican immigrants during the tumultuous thirties.[81]

Through the prism of gender, Cynthia Orozco provides fresh insights into the day-to-day operations and goals of the organization. From LULAC's inception, women participated in numerous grass-roots service projects. In Houston, for example, women raised "monies for milk, eye glasses, Christmas toys, and baby clothes." Orozco inscribes LULAC women's activism as concrete responses to community needs.[82] Their voluntarist politics indicates that there existed less of a social distance between immigrants and citizens as well as between workers and merchants than the rhetoric of LULAC would lead us to believe. LULAC women were bridge people simultaneously seeking to meet the material needs of newcomers and neighbors while engaging in direct action for civil rights.

Before World War II, Mexican middle-class women in Texas demonstrated their civic leadership as LULAC members and leaders. In 1937, Alicia Dickerson Montemayor from Laredo, Texas, became the first woman elected to the office of Second Vice-President General of LULAC, "the first . . . to hold an office not specifically designated for women." That same year, she published an article in *LULAC News* entitled "Son Muy Hombres," a distinctly feminist critique of Latin male privilege. In her words, "I was a very controversial person."[83]

In New Mexico, two remarkable women, Adelina Otero Warren and Concha Ortiz y Pino, ran for political office. In 1922, Adelina Otero Warren ran as a Republican candidate for the U.S. House of Representatives, the first New Mexican woman and the first Hispana to run for national office. María Adelina Isabel Emilia (Nina) Otero was born in 1881, a member of the "Spanish" land-owning gentry of New Mexico. The extended Otero family was well known for its political clout, with Nina's second cousin Miguel Otero serving as territorial governor from 1897 to 1906. Well educated in Santa Fe and St. Louis, Otero Warren ventured to New York City for a brief stint in settlement work.[84] From 1915 to 1920, the strong-willed, independent Hispana proved a tireless crusader for women's suffrage. As she wrote Alice Paul of the Congressional Union and later the National Women's Party, "[I] will take a stand and a firm one whenever necessary for I am with you now and always." Paul, in turn, recognized her contributions as an influential state lobbyist and in a telegram credited Otero Warren's "splendid leadership" in securing New Mexico's ratification of the Nineteenth Amendment.[85] According to her biographer Charlotte Whaley, Nina Otero Warren had become "a heroine in Santa Fe, an intelligent, resourceful woman with a strong sense of justice." She continued, "More than any other New Mexican woman of that time, she had become a role model for both Anglo and Hispanic women in the state."[86]

Emboldened perhaps by her celebrity and her election as Su-

perintendent of Schools for Santa Fe County in 1918, Otero Warren decided to run for Congress. Winning the Republican primary,
she proved an astute politician. She penned press releases detailing
her reform record on education, temperance, and social welfare. A
charismatic campaigner, she spoke of her political ideals and her
connection to the land, beginning one speech with the claim that
"her people had founded the state." Since her Democratic opponent was an Anglo male, she may have used the phrase "my people"
to encompass all Hispanos, not just her own genealogical family.[87]
Certainly her gender and her "Spanish noble heritage" made good
newspaper copy. One reporter went so far as to attribute her red
hair and freckles to her German Visigoth ancestors who had centuries before settled in Spain. Otero Warren ran on a record of hon-
esty, integrity, and fair play; but her cousin, the former governor
Miguel Otero, publicly revealed that she had lied about her personal life. Adelina Otero Warren was not the widow of Rawson
Warren, but a divorcee. Historian Elizabeth Salas contends that
news of her divorce (which had occurred twelve years earlier)
helped tip the scales in favor of her Democratic opponent. Defeated by less than 10,000 votes, Otero Warren never again sought
an elected state or national office.[88]

Adelina Otero Warren, however, remained a force in New Mexico education. Although she opposed bilingual education, she did
prove a vocal critic of boarding schools for Native Americans. She
was also concerned with cultural conservation, particularly with regard to Hispano arts and crafts. Literary critic Diana Rebolledo
considers Otero Warren's *Old Spain and the Southwest* (1936) a
narrative of resistance to the Anglocization of New Mexico. Her ro-
mantization of a halcyon Spanish colonial past fits within the politics of identity prevalent among Hispanos in New Mexico and
southern Colorado.[89] I suggest that Adelina Otero Warren traversed at least two worlds on two levels as she negotiated across Hispanic and Euro-American New Mexico as well as across her own
subjectivities as both rural aristocrat and modern feminist.

In 1936, the first Spanish-speaking woman took her seat as a
member of a state legislature. At age twenty-six, Concha Ortiz
y Pino from Galisteo, New Mexico, also became the youngest
woman elected to state office. During her six years as representative from Santa Fe County, the media frequently depicted her
as a "glamour girl," but on the floor she proved a conscientious
and shrewd legislator. With confidence, style, and on occasion a
tart tongue, she garnered respect and, during her last term, Ortiz y

Pino, a Democrat, served as the majority whip in the New Mexico House of Representatives.[90]

Like Adelina Otero Warren, Concha Ortiz y Pino traced her ancestry to Spanish nobility; her family had participated in electoral politics since territorial days.[91] Ortiz y Pino, however, ran for office out of a sense of noblesse oblige rather than a desire for progressive social reform. During our interview, she recalled how, at Christmas, her father José would reiterate how the family's position in the region ("the many more gifts the Lord has given you") required them to be generous and oversee the welfare of their neighbors and employees. With shades of a nineteenth-century seigneurial world view, she echoed her father's words:

> The family responsibility is always this: first to your government, secondly your loyalty . . . to your church; and thirdly, an unwavering family loyalty and care of those who work for you. You must take care of their material needs, their spiritual needs, and their medical needs.[92]

She further related that "after the Americans moved in and took over New Mexico," her great-grandfather Don Nicolás vowed that "a Pino or related Ortiz would stand for election to the state assembly in every generation." Her father had served in the state house and Concha followed in his footsteps. When I asked her why she, rather than her brother Frank, had been selected, she laughingly explained, "Oh, they wouldn't throw him to the wolves. They picked me."[93]

I would argue that her family recognized early her aptitude for public life. "My father spent 10 years teaching me the rules of the house. . . . Every afternoon when [he] was in the state legislature, I had to go and sit there after school."[94] In 1930, at the age of eighteen, she organized local villagers at Galisteo into a vocational school for the promotion of traditional Hispano crafts, such as blankets, leather goods, and furniture. This artisan venture became so successful that it attracted the attention of the State Department of Vocational Education and, more important, provided a viable livelihood for Galisteo families hard hit by the Depression. In 1936, she switched her party affiliation from Republican to Democrat and with a chauffeur as her chaperone, she campaigned for both a statehouse seat and FDR's re-election.[95]

Although a Roosevelt Democrat, Ortiz y Pino took a dim view of several measures identified with the New Deal. She voted against ratifying a proposed constitutional amendment prohibiting child labor, voted against state regulation of hours for working women, and proclaimed publicly that relief agencies did little more than provide jobs for "high salaried Lady Bountiful social workers."[96] She has never been identified as a feminist,[97] but, as a legislator, she did introduce the first bill that would have made women eligible to serve on juries. Her measure failed and it was not until 1969 that New Mexico permitted women to sit in a jury box. She also pushed for a civil service merit system that would protect women workers at the capital from bosses who expected sexual favors as a condition of employment.[98] Her most notable legislative accomplishment involved introducing and securing the passage of a bill requiring the teaching of Spanish in New Mexico's grade schools. An early proponent of bilingual education, she resigned her membership in LULAC when it publicly opposed her measure.[99]

By the time Ortiz y Pino had finished her last term in 1942, she had also graduated from the University of New Mexico and a year later married her favorite professor, Victor Kleven. For sixty years, Concha Ortiz Y Pino de Kleven has been a force to be reckoned with in New Mexico. Called "the most powerful woman" in the state, she has been a steadfast advocate for abused children and people with disabilities. In 1966, the *Albuquerque Journal* gave her credit for a state law that required public buildings to "be accessible to the disabled." She later remarked that Albuquerque commissioners called her "the Mother of the Ramps."[100] Concha Ortiz y Pino de Kleven, moreover, is well known as a doyenne of the arts and humanities. An ageless aristocrat, she views a good life as one devoted to public service. Reflecting a voluntarist politics rooted in noblesse oblige, Concha Ortiz y Pino exemplifies a quintessential *Doña* in New Mexico civic life, a leader with an entitled sense of place and heritage. "I don't need a costume . . . I know who I am."[101]

Public political leadership among Spanish-speaking women was not limited to an aspiring middle class or an elite gentry. Organized by Luisa Moreno, El Congreso de Pueblos de Hablan Española (the Spanish-speaking Peoples Congress) represented the hopes and dreams of many working-class Latinos. By the late 1930s, Moreno was cognizant of both the strength of local institutions and the distance between immigrants and citizens as she endeavored to bring together community networks under the umbrella of a national civil rights congress.[102]

El Congreso de Pueblos de Hablan Española was the first national civil rights assembly for Latinos in the United States. The approximately 1,000 to 1,500 delegates representing over 120 organizations assembled in Los Angeles on April 28 to 30, 1939, to address issues of jobs, housing, education, health, and immigrant rights. Luisa Moreno, its driving force, drew on her contacts with Latino labor unions, mutual aid societies, and other grass-roots groups to ensure a truly national conference. Los Angeles activists Josefina Fierro de Bright, Eduardo Quevedo, and Bert Corona would also assume leadership roles in El Congreso. Although the majority of the delegates hailed from California and the Southwest, women and men traveled from such distances as Montana, Illinois, New York, and Florida to attend the convention. According to historian Rudy Acuña, El Congreso "was broadly based: workers, politicians, youth, educators—people from all walks of life."[103]

Congreso delegates drafted a comprehensive platform. They called for an end to segregation in public facilities, housing, education, and employment and to discrimination in the disbursement of public assistance. El Congreso endorsed the rights of immigrants to live and work in the United States without fear of deportation. While encouraging immigrants to become citizens, delegates did not advocate assimilation, but instead emphasized the importance of preserving Latino cultures and called on universities to create departments in Latino Studies. Despite the promise of the first convention, a national network of local branches never developed and red-baiting would later take its toll among fledgling chapters in California.[104]

Scholars have debated the extent to which the United States Communist Party actually influenced the Spanish-speaking Peoples Congress.[105] Historian George Sánchez offers the following insights:

> By not excluding Communists from their ranks—both Anglos and Latinos—El Congreso proved to be an inclusive organization, but not one "captured" by any outside group. In fact, the leadership of El Congreso, though clearly a product of labor and left organizations, prided itself on being able to appeal to *all* Latinos, regardless of political affiliation.[106]

El Congreso brought together two dynamic women—Luisa Moreno and Josefina Fierro de Bright—whose life-long friendship was forged in the fire of community organizing. Born into an elite Guatemalan family, Luisa Moreno rejected her patrimony early and as a teenager ventured to Mexico City where she worked as a journalist and pursued her talents as a poet. In 1927, at the age of twenty-one, she married artist Angel De León and the couple immigrated to New York City in 1928. A few months later, Luisa gave birth to her only child, a daughter Mytyl. With the onset of the Great Depression, Moreno, struggling to support her infant daughter and unemployed husband, bent over a sewing machine in Spanish Harlem. She organized her compañeras into a Latina garment workers' union. Her talents did not go unnoticed. In 1935, the American Federation of Labor hired her as a professional organizer. Leaving an abusive husband, Luisa, with Mytyl in tow, boarded a bus for Florida where she would unionize African-American and Latina cigar rollers. Within two years, she joined the CIO and in 1938 she became an UCAPAWA representative.[107]

A native of Mexicali, Josefina Fierro descended from a line of rebellious women. Her grandmother and mother were Magónistas, followers of socialist leader Juan Flores Magón and Partido Liberal Mexicano. Josefina remembered her mother Josefa teaching her important lessons at an early age. "She taught us to fight discrimination. She taught us to organize, what community was . . . that it wasn't a shame to be a Mexican."[108] Growing up in Los Angeles and the San Joaquin Valley, Josefina and her brother helped her mother prepare and sell food. At one point, Josefa supported her family by converting a trailer into a boardinghouse and restaurant on wheels to better serve their migrant clientele. While a student at UCLA, Josefina Fierro met Hollywood writer John Bright at the local cabaret where her tia sang.[109] After their marriage, she became a community organizer in East Los Angeles drawing on her celebrity connections to raise funds for barrio causes. While Luisa Moreno took the lead in organizing the 1939 national meeting of El Congreso, Josefina Fierro de Bright proved instrumental in buoying the day-to-day operations of the fragile southern California chapters. Recalling her East Los Angeles days and Hollywood nights, she simply stated, "I was always sleepy."[110]

Moreno and Fierro de Bright believed in the dignity of the common person and the importance of grass-roots networks, reciprocity, and self-help. As Josefina commented, "Movie stars such as Anthony Quinn, Dolores Del Rio, and John Wayne contributed money, not because they were reds . . . but because they were helping Mexicans help themselves."[111] The two women also shared an awareness of the positionality of women in U.S. Latino communities. As we will see in the following chapter, El Congreso created a woman's committee and a woman's platform, a platform that expressly recognized the "double discrimination" facing Mexican women. Josefina Fierro de Bright put it this way: "We had women's problems that were very deep . . . discrimination in jobs . . . migratory problems . . . schooling. No, we didn't have a Lib Movement so we didn't think in terms of what women's roles were—we just did it and it worked."[112]

Given their beliefs and backgrounds, it is not surprising that both women cared passionately about immigrant rights. Speaking before the 1940 conference of the American Committee for the Protection of the Foreign Born (ACPFB), Luisa Moreno used the metaphor "Caravan of Sorrow" to delineate the lives of Mexican migrant workers. A portion of her presentation follows:

> Long before *The Grapes of Wrath* had ripened in California's vineyards a people lived on highways, under trees or tents, in shacks or railroad sections, picking crops—cottons, fruits, vegetables, cultivating sugar beets, building railroads and dams, making barren land fertile for new crops and greater riches.
>
> These people are not aliens. They have contributed their endurance, sacrifices, youth and labor to the Southwest. Indirectly, they have paid more taxes than all the stockholders of California's industrialized agriculture, the sugar companies and the large cotton interests, that operate or have operated with the labor of Mexican workers.[113]

Her talk made such an impact that the ACPFB reprinted selected sections in a pamphlet entitled "Non Citizen Americans of the South West." In southern California, Congreso members organized against anti-immigrant legislation introduced at the state capital. A caravan of "20,000 people by Fierro's estimate" arrived in Sacramento to voice their concerns. Governor Culbert Olson listened as he vetoed a bill that would have suspended relief to immigrants.[114] While not to diminish the desegregation and voter registration efforts of such organizations as LULAC, American G.I. Forum, ANMA, or CSO,[115] I would argue that a direct link exists between the platform endorsed by El Congreso and many of the goals articulated within the Chicano Student Movement.

Luisa Moreno and Josefina Fierro de Bright were foremothers to women who would come to call themselves Chicana/Latina feminists. Although political gains were made during the 1950s, especially with regard to desegregation, much work remained. The Mexican neighborhood of Chávez Ravine, for example, was leveled to make way for Dodger Stadium. According to urban planner Victor Becerra, "Aside from displacing 7,500 people, destroying some 900 homes, and costing taxpayers approximately $5,000,000, the arrival of baseball to Los Angeles didn't change things much."[116] In 1960, the per capita income for Mexicans in the Southwest averaged "$968 compared to $2,047 for Anglos." Furthermore, the median years of schooling completed by Mexican workers in 1960 was only 7.1 years in comparison to 12.1 for Euro-Americans and 9.0 for "non-whites."[117] As part of global student movements of the late 1960s, Mexican American youth joined together to address continuing problems of discrimination, especially in education and political representation. They transformed a pejorative barrio term "Chicano" into a symbol of pride. "Chicano/a" implies a commitment social justice and social change.

Self-help, reciprocity, and commadrazgo are woven through the narratives of Mexican-American women. With pickets, baskets, and ballots, they created tapestries of resistance. Representing a range of ideologies from proletarian to seigneurial, they addressed injustice and served their communities. During the 1960s and 1970s, women in the Chicano Student Movement would draw on these legacies to form a Chicana generation that reclaimed, reinscribed, and transformed political and cultural subjectivities—with feminism as the contested *frontera*.

5

La Nueva Chicana:
Women and the Movement

❖

Rise Up! To Woman
Rise up! Rise up to life, to activity, to
the beauty of truly living; but rise up radiant
and powerful, beautiful with qualities, splendid
with virtues, strong with energies.[1]

THIS poem was written not during the heyday of the Chicano
Movement, but in 1910 by Tejana socialist labor leader and political activist Sara Estela Ramírez.[2] She would not live to participate
in El Primer Congreso Mexicanista held the following year. Ramírez's ideas, however, would resonate in the words of her compañeras. Composed of South Texas residents, this Congreso was
the first civil rights assembly among Spanish-speaking people in
the United States. With delegates representing community organizations and interests from both sides of the border, its platform addressed discrimination, land loss, and lynching. Women delegates,
such as Jovita Idar, Soldedad Peña, and Hortensia Moncaya, spoke
to the concerns of Tejanos and Mexicanos. Regarding education,
Soledad Peña referred to "our duty . . . to educate woman; to in-
struct her and to . . . give her due respect." Out of this congress
arose a women's organization, Liga Femenil Mexicanista, and for a
short period two sisters Andrea and Teresa Villareal published a
newspaper, *La Mujer Moderna*.[3]

As the poem shows, strands of feminist ideology or incipient
feminist ideology can be located at various junctures in the history

99

of Mexican women in the United States. Feminism has taken many forms; however, this chapter focuses primarily on women's participation in the Chicano Student Movement "el movimiento" from 1968 to the present. The participation of "La Nueva Chicana" (to quote poet Viola Correa) in welfare rights, immigrant services and advocacy, sterilization suits, community organizations, La Raza Unida, campus activism, and literature has been reduced to a cursory discussion of sexism within the movimiento by the authors of the leading monographs on the Chicano Movement. Perhaps as frustrating, survey texts and relevant specialized monographs in U.S. Women's history overlook Chicana feminism.[4] Although scholars recognize the 1960s and 1970s as the era of the modern feminist movement, they have left Chicanas out of their stories. Countering these chilling silences, a growing body of scholarly studies and literary works offer eloquent testimonies of Chicana feminist thought inside and outside the academy. In this chapter, I will draw on them as well as archival materials and oral interviews to outline the ideological parameters of student politics and emphasize the ways in which women negotiated the terrains of nationalism and feminism, paying special attention to the iconography of the movimiento and the ways in which imagery from the Aztec world and the Mexican Revolution reinforced and challenged traditional notions of gender. The links between Marxism and feminism as well as debates over the double standard and intermarriage are examined. The chapter also discusses women's community organizing, the emergence of Chicana lesbian consciousness, and the call for third-world coalitions.

The titles of two recent publications illuminate a basic dialectical tension among Chicana feminists—*Building with Our Hands* and *Infinite Divisions*. This chapter offers a brief glimpse into the ways in which Chicanas have articulated a community-centered consciousness and a recognition of differences as they live amid the "swirls of cultural contradictions."[5] I would like to start with the development of one of the basic themes of Chicana feminism, a premise expressed by Tejana activist Rosie Castro. "We have practiced a different kind of leadership, a leadership that empowers *others*, not a hierarchical kind of leadership." Elizabeth (Betita) Martínez echoed, "The leadership that empowers others is the leadership we need."[6]

Over fifty years ago, Luisa Moreno and Josefina Fierro de Bright provided this type of leadership. The second California convention of El Congreso de Pueblos de Hablan Española, held in

December 1939, passed a prescient resolution with regard to working-class women. A portion follows:

> Whereas: The Mexican woman, who for centuries had suffered oppression, has the responsibility for raising her children and for caring for the home, and even that of earning a livelihood of herself and her family, and since in this country, she suffers a double discrimination as a woman and as a Mexican.
>
> Be It Resolved: That the Congress carry out a program of . . . education of the Mexican woman, concerning home problems . . . that it [a Women's Committee] support and work for women's equality, so that she may receive equal wages, enjoy the same rights as men in social, economic, and civil liberties, and use her vote for the defense of the Mexican and Spanish American people, and of American democracy.[7]

As I related in an earlier book, the women rank and file of UCA-PAWA gained heightened self-esteem along with an awareness of gender issues as the result of union activism. Sociologist Myra Marx Ferree has argued that women wage earners "are significantly more feminist" than housewives and that "the effect of employment is to place women into a social context which encourages feminist ideas."[8] Consistent with Ferree's findings, women labor activists, while perhaps not political feminists, certainly appeared attuned to sex discrimination within the canneries. Luisa Moreno recalled that during World War II in "Cal San negotiations a woman . . . member of the negotiating committee remarked: 'Females includes the whole animal kingdom. We want to be referred [to] as WOMEN. That remained henceforth in every contract.'"[9] In addition, southern California cannery workers demanded management-financed day care and an end to the piece-rate scale. UCA-PAWA's women members, in general, developed a job-oriented feminism; that is, they sought equality with men regarding pay and seniority *and* they demanded benefits that specifically addressed women's needs, such as maternity leave and day care.[10]

For some women, such as Julia Luna Mount, UCAPAWA served as a stepping-stone to life-long involvement in labor and civil rights issues. As a teenager, Julia Luna began working at Cal San and joined in the Workers' Alliance. She served as a rank-and-file leader during the 1939 Cal San strike and took part in El Congreso. She later left food processing to earn higher wages at McDonnell-Douglas, but was forced to quit because of unrelenting sexual harassment. She married George Mount, an organizer for the United Auto Workers, and continued her activist path. After World War II, she led union drives among Los Angeles hospital workers. In addition, she and her sister, Celia, both former cannery workers, were active in Asociación Nacional México-Americana (ANMA), a civil rights organization that emerged from the labor movement during the 1950s. Julia Luna Mount has run for political office several times under the banner of the Peace and Freedom Party. She was also well known in the southern California nuclear freeze movement.[11]

Whether through *commadrazgo*, *mutualistas*, labor unions, or political organizing, some women recognized what they had in common as women. Tensions *between* women, however, form a second theme of Chicana feminist history. This dialectic, often expressed as a conflict between personal liberation or family first, had emerged well before the nationalist student movement of the 1960s and 1970s. Examples abound of both paths. In Texas, Sara Estela Ramírez had encouraged women to "Rise Up!" and María Hernández did, indeed, take action—in concert with her husband. In 1929, Hernández and her husband Pedro formed Orden Caballeros de America, a *mutualista* that provided social services to the barrio as well as served as a springboard to political organizing. An activist for over fifty years, she also participated in school desegregation cases and, as a senior citizen, was an advocate of La Raza Unida, the Chicano third party. Hernández believed strongly in women's familial responsibilities as she elaborated in an essay published in 1945. As noted by Cynthia Orozco, Hernández adhered to the idea that "the domestic sphere was maintained to be the foundation of society and mothers the authority figures who molded nations." María Hernández has also been characterized as "a strong feminist," who emphasized "the importance of family unity" and "the strength of men and women working together." Hernández herself described her husband has "enlightened, committed, and liberated in every way, and who has never done anything but to encourage her participation in the community."[12] This statement resonates among contemporary heterosexual Chicana feminists. The phrase "My boyfriend/husband—he really is a feminist" represents a mantra of justification and affirmation.[13]

The birth of the Chicano student movement occurred in 1967 as Mexican Americans formed their own organizations on college campuses. Their numbers were relatively small; one survey revealed that the *cumulative* undergraduate enrollment for seven southwestern colleges included only 3,227 Mexican Americans (2,126 men, 1,101 women).[14] Yet, in May 1967, young adults gathered at Loyola University in Los Angeles and formed the United Mexican American Students (UMAS). In addition to holding fundraisers for the United Farm Workers and other community causes, this group sought to make alliances with local chapters of the Black Student Union (BSU) and Students for a Democratic Society (SDS). A year later, as a result of student pressure, the first Chicano Studies program was founded at California State University, Los Angeles.[15]

Activism was not limited to college campuses. A group of high school teens, including student council officers, circulated petitions urging the school board to take concrete measures to improve the quality of secondary education in East Los Angeles. Board members politely received the petitions and then discarded them. As a result, in March 1968, over 10,000 youngsters at five area schools (Roosevelt, Wilson, Lincoln, Garfield, and Belmont) walked out. Staging the largest student walkout in the history of the United States, the young leaders had now captured the attention of the board. They demanded a revised curriculum to include Mexican/Chicano history and culture; the recruitment of more Mexican-American teachers; an end to the tracking of Chicano students into vocational education; and the removal of racist teachers. They also desired smaller classes and upgraded libraries. Vicky Castro recalled that issues ranged "from better food all the way to . . . we want to go to college."[16] Another student who walked out, artist Patissi Valdez, succinctly related the attitude of her home economics teacher as an example of the lessons taught at her school:

> She would say . . . "You little Mexicans, you better learn and pay attention. This class is very important because . . . most of you are going to be cooking and cleaning for other people."[17]

Students, moreover, had few Mexican-American role models as "only 2.7 percent of the teachers . . . had Spanish surnames." One of these educators, Sal Castro, joined the protesters; he could not in good conscience remain inside the walls of Abraham Lincoln High School. Perhaps with a twinge of nostalgia, he remembered the walkouts (or blowouts) as: "Kids out in the streets with their heads held high. With dignity. It was beautiful to be a Chicano that day."[18]

The East Los Angeles "blowouts" lasted over a week. One *Los Angeles Times* reporter referred to the protests as "the birth of brown power." The media keyed in on banners carried by students: "Chicano Power," "Viva La Raza," and "Viva La Revolución."[19] The LAPD overreacted at Roosevelt and Belmont high schools, chasing and bludgeoning teens. Describing the scene, Mita Cuarón declared, "It didn't match the thing we were doing. We didn't commit a crime. We were protesting." The blowouts did initiate reform in Los Angeles schools. "We were very successful at informing the public about how serious the conditions were," reflected Paula Cristonomo. In fact, Senator Robert Kennedy met with the students and sent a telegram of support.[20] Walkouts in Mexican schools followed in such disparate cities as Denver, Phoenix, and San Antonio. Such militancy was not confined to the Southwest. Writer Ana Castillo, a native of Chicago, recalled her own adolescent activism. "I went downtown and rallied around City Hall along with hundreds of other youth screaming 'Viva La Raza' and 'Chicano Power!' until we were hoarse." Demonstrations also proliferated on college campuses. A coalition of students of color orchestrated the 1969 Third World Strike at Berkeley. Facing police batons and arrests, these students called for the creation of a third college dedicated to people of color and run by the community. A Department of Ethnic Studies on the Berkeley campus was the concrete result.[21]

In March 1969, Colorado community activist Corky Gonzales and the Crusade for Social Justice hosted the National Chicano Youth Liberation Conference at Denver, Colorado. This conference offered a potent nationalist vision linking the Aztec past to a Chicano future. The concept of Aztlán as the mythic Aztec homeland reborn in a Chicano nation resonated among the audience, over 1,500 strong.[22] A former organizer with the Student Non-Violent Coordinating Committee (SNCC) in Alabama, María Varela gave eloquent testimony to the transformative power of this gathering:

> "Conference" is a poor word to describe those five days. . . . It was in reality a fiesta: days of celebrating what sings in the blood of a people who, taught to believe they are ugly, discover the true beauty in their souls . . . Coca, Cola . . . Breck Shampoo, the Playboy Bunny, the Arrow Shirt man, the Marlboro heroes are lies. "We are beautiful . . ." [T]his affirmation grew into a *grito*, a roar among the people gathered in the auditorium.[23]

A month later at a youth conference at the University of California, Santa Barbara, from which MECHA (El Movimiento Estudantil Chicano de Aztlán) was officially launched, students stressed the importance of applying their education for the benefit of their communities.[24] Echoing that "man is never closer to his true self as when he is closer to his community," El Plan de Santa Barbara expounded that:

> *Chicanismo* draws its faith and strength from two main sources: from the just struggle of our people and from an objective analysis of our community's strategic needs. . . . Chicanos recognize the central importance of institutions of our higher learning to modern progress, in this case, to the development of our community.[25]

Whether one views the Chicano Student Movement as a political quest or as a nationalist struggle, one cannot subsume its identity under the rubric of "Me, too." Although there were a few connections to African-American civil rights groups, with SNCC veterans Betita Martínez and María Varela bringing their organizing skills and experiences to the Southwest, the Chicano Movement was very much its own entity with its own genesis. However, in U.S. history textbooks, Mexicans are typically relegated to the end of the book and pictured as either followers of Cesar Chávez or student activists emulating African Americans. It was not that they wanted a piece of the "American pie," they wanted the freedom to bake their own pan dulce.[26]

Situating one's politics, indeed one's very life, toward community empowerment was a given among Chicano student activists. Forging bonds of community with one another and, most important, with the worlds they literally or metaphorically left behind crossed ideological borders as they created diffuse and at times competing organizations. Whether one assumed the mantle of cultural nationalism (working only for Chicanos) or longed for third-world liberation, sustaining connections to a world outside the university proved crucial. Students led food drives for the United Farm Workers; protested the Vietnam War; offered tutorial programs for barrio youth; organized for immigrant rights, volunteered for La Raza Unida, the Chicano third party; established health clinics; and launched Chicano Studies programs.[27] Success was measured in terms of social justice, not material wealth.

Aztec motifs dominated the iconography of the Chicano Student Movement. Young militants adopted Aztec imagery and heritage as their own. The image of the "warrior" struck a cord with "its ferociously macho imagery." While noting that the idea of a Brown nation (or Aztlán) offered a "taste of self-respect," long-time community activist Betita Martínez argues that "merely as a symbol the concept of *Aztlán* encourages the association of machismo with domination."[28]

What roles could women play in this hagiography of a pre-Columbian past? Think about representations of Aztec life that you have seen: the familiar murals in Mexican restaurants and calendar art. Light-complected women dressed in translucent gowns held in the arms of muscular bronze men arrayed in gold and feathers. The spectacular mural at the El Cerezo Restaurant in El Paso, Texas, serves an example. Against the backdrop of mountain scenery, an Aztec warrior bends his head in sorrow over a supine vestal virgin. This feminine icon (dead, no less) can hardly be construed as an empowering symbol for Chicanas.

Worse yet are traditional notions of La Malinche. Born of Aztec nobility, La Malinche (Malintzin Tenépal) was sold by her mother into a state of slavery at the age of eight. Six years later, she was given to Hernán Cortés who soon made use of her linguistic and diplomatic skills. La Malinche would also bear him a son. Viewed as a traitor to her people, La Malinche remains "the Mexican Eve." In this vein, Carlos Fuentes argues that Mexican women are twice cursed—they bear both the sin of Eve and the sin of La Malinche.[29] Today in Mexico the term Malinchismo means selling out to foreigners. A popular Mexican ballad of the mid-1970s bore the title "The Curse of La Malinche." A excerpt follows:

> The curse of offering foreigners
> Our faith, our culture,
> Our bread, our money,
> Remains with us.
>   . . .
> Oh, curse of Malinche!
> Sickness of the present
> When will you leave my country
> When will you free my people?[30]

Given these symbolic meanings, one of the first tasks Chicana feminists faced was that of revising the image of La Malinche. Adelaida Del Castillo's pathbreaking 1977 article provided a new perspective by considering Malinche's captivity, her age, and most important her conversion to Christianity. What emerges from Del Castillo's account is a gifted young linguist who lived on the margins and made decisions within the borders of her world.[31] Lucha Corpi offers a sympathetic and tragic portrait of Malinche in "Marina Mother."

> They made her of the softest clay
> and dried under the rays of the tropical sun.
> With the blood of a tender lamb
> her name was written by the elders
> on the bark of that tree
> as old as they.
>
> Steeped in tradition, mystic
> and mute she was sold—
> from hand to hand, night to night,
> denied and desecrated, waiting for the dawn
> and for the owl's song
> that would never come;
> her womb sacked of its fruit,
> her soul thinned to a handful of dust.
>
> You no longer loved her, the elders denied her,
> and the child who cried out to her "mamá!"
> grew up and called her "whore."[32]

Other Chicana writers envision Malintzin as a role model. Cordelia Candelaria considers her "a prototypical Chicana feminist," a woman of "intelligence, initiative . . . and leadership." With similar intent, poet Carmen Tafolla casts her as a woman who dared to dream.[33]

In this way, picking up the pen for Chicanas became a "political act." As poet Naomi Quiñonez explained, "With poetry, I could encourage, reaffirm, and mirror efforts toward social change. I wrote poetry while at antiwar rallies, during class discussions about capitalism, at Cinco de Mayo celebrations . . . and at Santana concerts." Quiñonez considered herself a cultural worker as she planned poetry readings at barrio community centers, parks, and churches as well as on college campuses.[34] Women also founded and edited newspapers—*El Grito* (Betita Martínez); *Encuentro Feminil* (Adelaida del Castillo and Ana Nieto Gómez); *Regeneración* (Francisca Flores); and *El Chicano* (Gloria Macias Harrison). Through their writings, Chicanas problematized and challenged prescribed gender roles at home (familial oligarchy); at school (the home economics track); and at meetings (the clean-up committee).[35]

From the early days of the student movement, women were not always satisfied with the rhetoric and praxis of their compañeros, but those who called for the introduction of women's issues on the collective agenda or for an end to gender-specific tasks (e. g., typing, cooking, cleaning up) were labeled "Women's Libbers" or *"aggringadas,"* and open ridicule had a chilling effect. At the 1969 Denver youth conference, after considerable discussion in the women's workshop, a Chicana facilitator reported to the conference as a whole: "It was the consensus of the group that the Chicana woman does not want to be liberated." "I felt this as quite a blow," penned conference participant Enriqueta Longauex y Vasquez soon after the event. "Then I understood why the statement had been made and I realized that going along with the feelings of the men at the convention was probably the best thing to do at the time."[36] Sonia López reported that some coeds proved reluctant to join campus Chicana organizations, like Hijas de Cuauhtemoc at Cal State, Long Beach, for fear of being labeled or rejected by men. In the words of Francisca Flores, "Women must learn to say what they think and feel, and free to state it without apologizing or prefacing every statement to reassure men that they are not competing with them."[37]

By 1971, in Houston, Texas, at La Conferencia de Mujeres Por La Raza, the first national Chicana conference, women spoke out with a distinctly feminist platform. The resolutions called for "free legal abortions and birth control in the Chicano community be provided and controlled by Chicanas." In addition, they called for higher education, for acknowledgment of the Catholic Church as an instrument of oppression, for companionate equalitarian marriage ("Marriage-Chicana Style"), and for child care arrangements to ensure women's involvement in the movement. Of the 600 women who attended the conference, as many as one-half disagreed with such radical resolutions and walked out. They believed that in-house problems should not overshadow community concerns. ["Our enemy is the gavacho not the macho."][38]

The Houston conference, however, did place feminism very visibly on the movement table. Chicano leaders either pretended it didn't exist or dismissed women's concerns. Corky González openly admitted he did not want Chicanas to become like the "frigid gringa."[39] He also issued a dire warning about the hidden *traditional* power of women:

> A woman who influences her old man only under the covers or when they are talking over the table, and when he goes in—if it's a bad idea—and argues for that, because he's strong enough to carry it through, is doing a disservice to *La Causa*.[40]

These attitudes could also be found in movement songs. In "The Female of Aztlán," women were told that "their responsibility is to love, work, pray, and help . . . the male is the leader, he is iron, not mush."[41] Ramón Gutiérrez accurately suggests that men "initially regarded the feminist critique as an assault on their Mexican cultural past, on their power, and by implication their virility." In fairness, some Chicanos envisioned themselves as placing women on bronze pedestals as Chicana queens or Aztec princesses.[42] The heartfelt (but inept) poem "Mi Amor" echoes familiar pedestal themes found in Mexican serenades sung a century ago. Two stanzas should suffice:

> To my eyes
> A Chicana is an exotic queen,
> She radiates a glow of exquisite
>   sheen.
>
> . . .
>
> She has style, she has class,
>   and she is aware,
> That her beauty is exceptional,
> So she walks tall and proud,
>   never forgetting
> That her beauty is traditional.[43]

Whether queen for a day or maid for a week, many Chicanas chafed at images or roles that placed them in the category of "traditional" helpmate.

In various venues from *Encuentro Femenil* to *Mademoiselle*, Chicanas passionately articulated their frustrations as women within the movement. "A lot of women were finding themselves unfulfilled in being just relegated to this position of beast of burden." They also questioned the double standard and blew the whistle on men who used political rhetoric for sexual conquest ["the guys that radicalize her off her pants"].[44] Feminists also problematized the madonna/whore dichotomy in Latino culture, considering it a holdover of colonialism. As one woman stated, "And if you're active in Raza Unida, you're suspected of being la mujer mala and in order to prove you're not, you have to live the life of a nun."[45] Ramón Gutiérrez would find within this Chicana feminist discourse a common thread, "a story that was rooted in the politicization of the body."[46]

Adelaida del Castillo succinctly declared that "Chicana feminism . . . recognized the worth and potentials of all women."[47] Chicanas, however, deliberately distanced themselves from Euro-American feminists. The pivotal anthology *This Bridge Called My Back,* edited by Cherríe Moraga and Gloria Anzaldúa, is replete with painful, searing tales of encounters between feminists of color and Anglo liberationists. Native American/Chicana writer Chrystos observed: "The lies, pretensions, the snobbery and cliquishness, the racism. . . . The terrifying & useless struggle to be accepted. . . . I left the women's movement utterly drained."[48] Perhaps the objectification of their racial subjectivities—the exotic othering of their experiences galled women of color the most. Jo Carrillo's poem "And Take Your Pictures with You" encapsulates resistance to such objectification. A portion follows:

> Our white sisters
> radical friends
> love to own pictures of us
> sitting at a factory machine
> wielding a machete
> in our bright bandanas
> . . .
> And when our white sisters
> radical friends see us
> in the flesh
> not as a picture they own,
> they are not quite as sure
> if
> they like us as much.
> We're not as happy as we look
> on
> their
> wall.[49]

Chicanas recoiled against the middle-class orientation of liberal feminists; the anti-male rhetoric of radicals; and the condescending, dismissive attitudes expressed by both.[50] Conflicts over gender or race, personal liberation or family first, did not stop the development of Chicana feminism. Caught between "maternal" and "paternal" movements, Ana Nieto Gómez declared:

> The Chicana feminist has been cautioned to wait to fight for her cause at a later time for fearing of dividing the Chicano movement. Also it has been recommended that she must melt into the melting pot of femaleness rather than divide the women's movement.[51]

At first these attitudes divided Chicanas themselves into two camps: feminists and loyalists. The loyalists believed that one should "stand by your man" and "have babies por la causa." They argued that Chicanas who needed "an identity" were *"vendidas"* or *"falsas."* Firing back, Nieto Gómez laid out the following scenario. "A girl may find that an open avenue to temporary status and distinction is to sleep with a noted 'Heavy.' . . . This is the traditional 'back door' open to all ambitious women."[52]

And then a third image also emerged—"La Adelita" or the *soldadera.* The idea of a strong, courageous woman garbed in the iconography of the Mexican Revolution was not a threatening image to Chicano nationalists; it implied that the woman fights beside her man and cares for his needs. Women in the Brown Berets, especially, were seen as contemporary *soldaderas.* Norma Cantú posits that Chicano nationalists did not embrace the more historically exact representation of the *soldadera* as a solider in her own right, but instead clung to the popular stereotype perpetuated by the Mexican cinema in which "Pedro Armendariz rides into the next battle [as] Dolores del Rio follows—on foot."[53]

But women, too, embraced this icon. The *soldadera* embodied a conflicted middle ground between loyalist and feminist, one that could be fiercely independent, yet strongly male-identified. The newspaper *El Rebozo* from San Antonio, Texas, "was written by women, put on by women, distributed by women, and was undertaken for the purpose of uniting our people to work for La Causa."[54] Yet even the name *El Rebozo* literally and figuratively wraps Chicana consciousness within the bosom of the *soldadera.* As the editors explained:

> El Rebozo—the traditional garment of the Mexican woman, with its many uses, symbolizes the three roles of the Chicana, portraying her as "la señorita," feminine yet humble; as "la revolu-

cionaria," ready to fight for "La Causa," and finally portraying the role of "La Madre" radiant with life.[55]

In a poem "Youth Mirror," Ana Nieto Gómez reminded Chicanas of Mexican women warriors, Juana Galla and Petra Ruiz, revolutionary commanders who had led both men and women. Nieto Gómez endeavored to link historical action with Chicana liberation.[56]

Feminists, however, frequently found themselves isolated as individuals. Referring to women's participation in La Raza Unida, Marta Cotera noted the costs involved for women who seemed too outspoken on gender issues. To stay in the Party, Cotera related, "These women must retreat into more conservative stands, and in effect, retreat from certain activities that deal specifically with the special social needs of women." If one dropped the mask of agreeability, there were consequences. Indeed, Ana Nieto Gómez would be silenced as she faced physical threats and was later fired from her position in Chicano Studies at California State University, Northridge. The "gender objectification" inherent in cultural nationalist ideology had painful, concrete manifestations for women, such as Nieto Gómez, who openly articulated a Chicana feminist vision. As literary theorist Angie Chabram-Dernersesian would later write, "Chicanas were denied cultural authenticity and independent self-affirmation."[57]

At times biting their lips, most Chicana feminists chose to remain involved in the Movimiento. Rhetorically, some contended that feminism was not divisive. During the early 1970s, Ana Nieto Gómez somewhat idealistically for the times wrote that the "feminist movement is a unified front made up of both men and women—a feminist can be a man as well as a woman—it is a group of people which advocates the end of women's oppression." Considering Chicana feminism as "part of the Chicano Movement," Adelaida del Castillo stated that "we have to work together in order to save ourselves. . . . As Chicanos we have the responsibility to look after each other."[58] Meanwhile, many feminists chose to focus their activities on women's concerns, such as defending the rights of welfare mothers and protesting the forced sterilizations of poor women. They created their own organizations, such as Comisión Femenil Mexicana and the Chicana Action Service Center. Feminists could be found at every pivotal event and in every major movement organization from the Chicana Moratorium to La Raza Unida to CASA.

In 1967, Alicia Escalante began to organize welfare mothers in

East Los Angeles. On public assistance herself, Escalante sought to create a safe Chicana space for both mutual support and political action. Through the meetings of what would become the Chicana Welfare Rights Organization, she nurtured "a non-threatening environment where women felt comfortable enough to talk about the private issues one only revealed with family members or comadres." Comparing notes on pregnancy (or the lack thereof), some came to the painful realization that they had been sterilized while giving birth at the USC/Los Angeles County Medical Center.[59]

With the assistance of Comisión Femenil Mexicana, an organization founded by young Chicana professionals, twelve women sterilized at the hospital filed suit. They were represented by two attorneys from the Los Angeles Center for Law and Justice, one of whom was Antonia Hernández, a Comisión member fresh out of law school at UCLA. Her Comisión amiga Ana Nieto Gómez would later recall in an interview with historian Virginia Espino that doctors "thought they were throwing . . . in a free service. They saw themselves as agents of the public, saving taxpayers money." During the trial in 1978, anthropologist Carlos Velez Ibañez testified that he had found ample evidence of eugenics-infused attitudes among area doctors, including some listed on the Medical Center staff. He related an informant's tale of how Mexicans were denied pain killers until they had signed the consent form for a tubal ligation.[60] Detailing the specifics of the case in a law review article, Antonia Hernández wrote:

> All of the victims or near victims belonged to a racial minority, were poor, and could readily understand the English language. Most were approached for sterilization surgery while under the duress of labor, drugged, and confined. All of them entered the Medical Center without any intent of becoming sterilized, and all were persistently solicited for the operation. Many of the women encountered doctors and nurses who were openly hostile to them because of their ethnicity or poverty status.[61]

In *Madrigal v. Quilligan* (1978), Judge Jesse Curtis ruled in favor of the defendants; that is, in favor of the Medical Center and its doctors. Judge Curtis believed that the sterilizations had resulted from miscommunication rather than malice and encouraged physicians to explain more carefully the consequences of a tubal ligation to Mexicana immigrants. In the words of his decision, "One can sympathize with them for their inability to communicate clearly, but one can hardly blame the doctors for relying on those indicia of consent which appeared unequivocal on their face and which are in constant use at the Medical Center."[62] It had taken a great deal of courage for the twelve Mexicanas to come forward and tell their stories. The judge, however, ignored their voices and relied instead on an "ideology of cultural differences" that absolved the physicians of "the legal and moral responsibility for their actions." Although the women had lost in a court of law, the adverse publicity generated from the case and a series of investigative articles appearing in the *Los Angeles Times* prompted the Medical Center to revisit its policies and procedures with regard to informed consent.[63]

Feminist commitment to community took other forms as well. Founded by Comisión Femenil member Francisca Flores, the Chicana Action Service Center in East Los Angeles was "one of the first antipoverty agencies exclusively serving barrio women." The center offered job training and placement with 50 percent of its clients single mothers under thirty. Emphasizing the importance of its programs, Ana Nieto Gómez wrote: "The low retention of Chicanas in the secondary schools, 70% dropout before the tenth grade inadvertently fails to make available to the Chicana even the traditionally oriented feminine curriculum, i.e., typing, business, English, shorthand." She further indicated that Chicanas who dropped out of college found themselves "at the same door step" as their sisters who left high school.[64] Age discrimination plagued center graduates over the age of thirty-five. Yolanda Nava asserted that 41 percent could not find employment. Even government training projects were closed to them because their placement would seem "improbable."[65]

Now an institution in southern California, the Chicana Action Service Center has helped thousands of women secure jobs and pioneered the development of placement networks with corporations and nonprofit agencies. An early industrial advisory board for the center represented a "class" bridge between Chicana professionals and their barrio sisters. For instance, Carmen Olguin of Pacific Telephone and Sally Martínez of Southern California Edison offered frequent career planning and job interview workshops.[66] Delivering direct services, lobbying lawmakers, and conducting policy-oriented research, Comisión members returned to their communities with their education. Their legacy endures.

Chicanas joined with their compañeros in protesting the Vietnam War. These activists were only were only too cognizant of the impact of the war on the lives of their kin and neighbors. Between

1961 and 1969, Mexican Americans in the Southwest represented 10 to 12 percent of the population but accounted for almost 20 percent of the casualities. Feminist scholar and poet Adaljiza Sosa Riddell wrote that "too many" Chicano veterans came home "wrapped como enchiladas in red, white, and blue."[67] Roosevelt High student and blowout activist Tanya Luna Mount (daughter of George and Julia Luna Mount) summed up the feelings of many by linking education in the barrio and the war in Vietnam. "Do you know why they [the Board of Education] has no money for us? Because of a war in Vietnam 10,000 miles away, that is killing Mexican-American boys—and for WHAT? We can't read, but we can die! Why?"[68]

During the Chicano Moratorium (August 29, 1970), women and children were fired on by police as they attended a peaceful protest rally in Laguna Park in East Los Angeles, a rally attended by over 20,000 people. In the aftermath of this "riot," a Los Angeles County deputy shot tear gas through a curtain hanging in the doorway of the Silver Dollar Bar. An unexploded tear gas canister killed K-MEX reporter and *Los Angeles Times* columnist Rubén Salazar as he sat in the bar having a beer and, no doubt, discussing the violence that had just transpired in the park. A televised inquest ruled Salazar's death an accident. The Chicano Moratorium shook the student movement to its core—sixty people had been wounded and three Chicanos, including Salazar, lost their lives. Remember that the Chicano Moratorium occurred only a few months after the killings at Kent State. Many Chicanos felt a mixture of moral outrage, anger, and disbelief. Moctezuma Esparza poignantly recalled, "It was a tremendous blow. Because we lost a certain heart, we lost a certain innocence of our ideals." He continued, "That we could engage the country . . . through the Bill of Rights, through the ideals that the country was supposedly founded on and that we could be killed."[69]

Disillusioned with the Vietnam War and Democratic Party promises, many Chicano activists called for the creation of their own third party. La Raza Unida (LRU) brought together politically engaged women across generations from María Hernández and Julia Luna Mount to Virginia Musquíz and Martha Cotera. Borne out of the high school walkouts in Crystal City, Texas, La Raza Unida became a political force in South Texas. With the prohibition of the poll tax, many small town Tejanos were eager to vote and the grassroots Chicano party captured their imagination and their hopes. In 1970, LRU candidates swept fifteen of the sixteen local races in

Crystal City, winning seats on the school board and the city council. Buoyed by local victories, LRU ran a slate for state office in
1972. The party, furthermore, spread outside the Long Star State
with strong showings in California and Colorado.[70]

Founded by José and Luz Gutiérrez, Virginia Musquíz, and
others, La Raza Unida became a vehicle for political self-determination, especially in South Texas. Tejanas played very visible roles
and several were elected to local office. Luz Gutiérrez asserted that
women made their position clear early on. "We actually had to walk
in to one of the meetings . . . and said, 'Hey we don't want to be the
tamale makers and . . . the busy bees. We really want to be part of
the decision-making process.'" Women also organized a separate
caucus within La Raza Unida (Mujeres Por La Raza) to promote
women's leadership on their own terms and not in the shadows of
LRU men. Acknowledging both empowerment and sexism within
its ranks, Martha Cotera also acknowledged LRU's nationalist
overlay. "When women came into the party, they fell under the
greater context of la familia de la raza."[71] Indeed, La Raza Unida
represented the zenith of cultural nationalism. Historian Ernesto
Chávez refers to LRU activism in Los Angeles as an "attempt to create Aztlán through the ballot box." Seeking to bridge the academy
and the community, Chicano college students provided the volunteer labor necessary to construct a base for California's La Raza
Unida Party. At a Chicano politics conference in northern California, Isabel Hernández railed against armchair activists. "It's going
to take a lot of work if we're going to make a political party," she
stated. "We have to stop the exploitation, we have to have personal
contact with people in the *barrio,* not sitting up in the colleges . . .
talking about revolution, but we also have to go to the *barrio* and
tell people where it's at."[72]

Although the party would fade by the mid-1970s, sociologist
Tatcho Mendiola contends that La Raza Unida "introduced to the
American public a brand new generation of leadership." As the re-
sult of the many threads of community organizing, including LRU,
"Within twenty years, there would be almost five thousand Latino
elected officials in the United States. Half would come from
Texas."[73]

Within a few years, cultural nationalism as the single, sustaining ideology waned and in its place emerged a renewed emphasis
on building class-based coalitions with Mexicanos, Latinos, and at
times third-world peoples. El Centro de Acción Autonoma-Her-

mandad General de Trabajadores (CASA) saw beyond the borders of Aztlán. Founded in 1968 by long-time civil rights activists Bert Corona and Chole Alatorre, CASA provided legal and social services for undocumented Mexicano/Latino immigrants. CASA, like El Congreso, believed in the rights of immigrants to work in the United States, join labor unions, and raise their families without fear. The historical link between the two organizations was not just ideological. At the age of twenty-one, Bert Corona had been a leader in El Congreso's youth division. David Gutiérrez refers to CASA as "the first Chicano-era organization to explore systematically the significance of the relationship between immigration, Chicano ethnicity, and the status of Mexican Americans in the United States."[74]

With an influx of student volunteers, CASA also conducted research on the contributions and exploitation of immigrants and lobbied against proposed anti-immigrant legislation. In 1975, Corona departed, turning over the reins to (in his words) "the young turks." Graduate students, undergraduates, and young professionals launched an ambitious agenda for what they considered a new and improved CASA. In addition to running the service center, they would educate the public through their newspaper, *Sin Fronteras,* organize workers into trade unions, and fortify themselves through research and study groups, openly Marxist in orientation. Corona would later remark that these leaders did not have the patience or experience necessary to build a constituency among Mexicano/Chicano workers. "They somehow believed that the workers would come on their own. . . . They thought the people would be attracted by the political line, by the rhetoric, and by the glamour."[75] Beset by problems that ranged from FBI surveillance to personal in-fighting, CASA disintegrated over the course of 1977 and 1978. Its slogans live on in the cadenced voices of contemporary Chicano/Chicana activists: "Obreros Unidos, Jamás Serán Vencidos" ["The Workers United Shall Never Be Defeated"] and "Somos Un Pueblo Sin Fronteras" [We Are One People Without Borders].[76]

CASA compañera, Magdalena Mora, provided a model of Chicana activism. A graduate student in history at UCLA, she not only wrote about trade union struggles but participated in them. She organized Mexicana/Chicana cannery workers in Richmond, California, at Tolteca Foods. As her professor Juan Gómez-Quiñones stated, "Ella hizo historia" [She made history]. Mora was also on the editorial staff of *Sin Fronteras*. An activist since her UMAS days as a high school student in San Jose, she died in 1981 of a brain tumor at the age of twenty-nine. Rudy Acuña eulogized her as a "beautiful fanatic" and "a movement person." In the words of Devra Weber, "Magdalena was a fighter. She saw and understood the potential in people. This motivated her political work and her struggle for a society in which that potential could be realized."[77] The informal credo of the Chicano student movement was to return to your community after your college education to help your people. Magdalena Mora never left.

Women in CASA were both inspired and frustrated. CASA could be considered "the signifier of Marxist thought in the Chicano Movement"; and perhaps, not surprisingly, women still had to fight for leadership roles. As Victor Becerra explained, "Women were not given positions of leadership; they took them." Considering themselves the revolutionary vanguard, CASA leaders required an all-encompassing commitment on the part of its members. People left school; people left jobs; their lives totally revolved around the organization. In some instances, even choice of dating partners was not an individual decision. Women did assume very visible public roles in CASA—as examples, Kathy Ledesma Ochoa and Isabel Rodríguez Chávez served as editors of *Sin Fronteras*. Chicanas in CASA, however, were expected to bear the brunt of "women's work" in planning fundraisers, selling tickets, and preparing food. The double standard prevailed. Hired by the ILGWU as an organizer, Patricia Vellanoweth discovered she was pregnant. Her compañeros berated her. "[Y]ou're supposed to be organizing, how could you get pregnant?" No one uttered a word to her husband.[78] Feminism did not appear on the collective agenda, although a few Chicanas openly identified themselves as feminists. Gender politics notwithstanding, involvement in CASA as organizers, writers, and service providers made an indelible imprint on their lives; many former Casistas remain active in advocacy efforts for immigrants. In the words of historian Marisela Chávez: "These women lived, breathed, and worked the movement."[79]

Surprisingly personal decisions, such as dating, marriage, and sexuality, became movement concerns whether one identified with cultural nationalism or Marxism or a some sort of combination or an in-between political space. The whole issue of interracial dating and marriage became hotly debated. In 1971, Velia García Hancock argued against this mixing on political grounds. It was not a question of "mingling of the bloods" given the nature of Mexican *mestizaje*, but rather that "intermarriage results in a weakening of ties and declining sense of responsibility and commitment to La Raza." These types of wholesale generalizations did little to promote communication. Love cannot be legislated. Furthermore, did marriage within La Raza always guarantee commitment to community empowerment? Many chroniclers and fighters for social justice, including slain journalist Rubén Salazar and poet/scholar/activist Adaljiza Sosa Riddell, intermarried. Marta Cotera addressed this issue in her 1977 collection of essays, *The Chicana Feminist*. "You have to be mature enough to respect people's choices. An individual who doesn't have freedom of choice cannot be liberated."[80]

From the earliest days of the Chicano Movement to current campus life, student activism has consistently included support for the United Farm Workers, an organization attracting volunteers representing diverse political orientations ranging from liberal to nationalist to Marxist. As we will see in the following chapter, Cesar Chávez and Dolores Huerta began to organize farm workers in the San Joaquin Valley in 1962. During the grape boycotts of the late 1960s and 1970s, the UFW was the most visible Chicano-identified organization, one that stressed a *mestizaje* of nationalism and coalition-building. Young people, whom David Gutiérrez refers to as "a multiethnic horde of idealistic . . . student volunteers," not only organized campus support groups but served in the fields. Many devoted their summers (and some dedicated a year or more) to the UFW as health workers, legal interns, boycott coordinators, media specialists, as well as organizers.[81]

In keeping with the spirit of El Plan de Santa Barbara, Chicano Studies departments encouraged undergraduate engagement with local issues. Teaching about the experiences of Mexicans in the United States was more than a classroom endeavor; experiential education became an essential component of a Chicano Studies curriculum. Internships with community-based organizations as part of field studies programs enabled young people to forge for themselves the links between the academy and the community.[82] Chicano Studies also provided a safe space, a home for Chicana and Chicano students, most of whom were first-generation college bound. In the poem "University Avenue," Pat Mora encourages undergraduates to draw on their cultural beliefs—*"cuentos whisper lessons en español"*—in coping with new terrain and new challenges.[83]

The problems faced by young women in MECHA or CASA during the 1970s continue to mark the development of Chicana feminism. Tugs-of-war persist over ideology, organizations, and gender relations. Some women turned to poetry as a means of articulating their opposition to machismo. In the classic "Machismo Is Part of Our Culture," Marcela Christine Lucero-Trujillo offered a look at the double standard:

> Hey, Chicano bossman
> don't tell me that machismo is part of our culture
> if you sleep
> and marry W.A.S.P
> You constantly remind me,
> me, your Chicana employee
> that machi-machi-machismo
> is part of our culture.
> I'm conditioned, you say,
> to bearing machismo
> which you only learned
> day before yesterday.
> At home you're no patrón
> your liberated gabacha
> has gotcha where
> she wants ya.
>
>   . . .
>
> Chicanismo through osmosis
> acquired in good does
> remind you
> to remind me
> that machi-machi-machismo
> is part of our culture.[84]

Chicana heterosexual undergraduates and graduate students perceived themselves as caught in a double bind in terms of suitable marriage partners among their compañeros. On one hand, many of their peers married white women and, in the minds of some Chicanos, these women were not Chicana enough—higher education had assimilated them to the point of no return. Noted Chicana scholar María Eugenia Matute-Bianchi recalled how in graduate school at Stanford a Chicano compañero had referred to her as "damaged and acculturated" because she desired a Ph.D.

Similar remarks could still be heard when I attended Stanford during the years 1977 to 1982. I recall the posturing of one graduate student who boasted that he never dated "college girls." "When I want a real woman, I go to the barrio in East San Jose and pick up a high school girl." He seemed genuinely taken aback when his remarks were challenged by the Chicanas who were within earshot.[85]

Claiming public space can be intertwined with sexuality. Chicana lesbians often found themselves isolated from Chicanos and heterosexual Chicanas. Referring to lesbians, in general, the following passage taken from the 1990 introduction to *Unequal Sisters* certainly speaks to Chicana lived experiences. "[L]esbians have struggled to create their own identities and to build their own communities in the midst of the most hostile environments. Carving a sense of sexual self amid such oppression was a courageous act of preservation, both personally and politically."[86] As Cherríe Moraga revealed, "My lesbianism is the avenue through which I have learned the most about silence and oppression."[87] In whisper campaigns and outright verbal assaults, Chicana lesbians were labeled *vendidas*, Malinches, and feminists. According to Alma García, "Feminists were labelled lesbians, and lesbians as feminists." At a Chicano activists' reunion, an organized conference held in San Antonio in 1989, a "joke" about Chicana lesbians found its way into the program.[88] Tejana literary critic María C. González poignantly recalled her own personal journey:

> I, too had been a product of a world that saw homosexuality as an immoral disease. My studies in feminism prepared me to accept my own lesbianism and also gave me an intellectual context for understanding my own resistance and fears about it.[89]

The works of Chicana lesbian writers Gloria Anzaldúa, Cherríe Moraga, Emma Pérez, Ana Castillo, Alicia Gaspar de Alba, Carla Trujillo, and many others bring out the pain and isolation, but, as important, their joys, self-respect, courage, and dignity. A sampling of this rich literature includes Gloria Anzaldúa's *Borderlands* and Cherríe Moraga's *Loving in the War Years*. The following four edited collections provide powerful testimonies of individual and collective struggles and affirmations: *This Bridge Called My Back*; *Making Face, Making Soul/Haciendo Caras*; *Compañeras: Latina Lesbians*; and *Chicana Lesbians: The Girls Our Mothers Warned Us About*. Equally important, these anthologies elucidate health issues, mixed race relationships, color consciousness, language, sexuality, self-representation, and historical constructions.[90] As critical works in feminist thought, *Bridge* and *Haciendo Caras* center the voices of lesbians of color as they dialogue with one another and the reader. Gloria Anzaldúa, furthermore, articulates a distinct lesbian fusion of private worlds and public spaces.

> As a *mestiza* I have no country, my homeland cast me out; yet all countries are mine because I am every woman's sister or potential lover. (As a lesbian I have no race, my own people disclaim me; but I am all races because there is the queer of me in all races.)[91]

The stories of Mexicana/Chicana lesbians have only begun to be told. At the University of Arizona, Yolanda Chávez Levya founded the first Latina lesbian archive. Alice Hom writes of "creating sites of resistance" among lesbians of color and her dissertation focuses on political coalitions that have been forged by lesbians in New York and Los Angeles.[92] These grass-roots networks offer hope in terms of building viable coalitions among women of color. Many more comparative studies are needed to bring together the voices of Latinas, indigenous women, Asian Americans, and African Americans, all women who live on the borderlands.

Chicana lesbians have painfully articulated the oppression they have experienced at the hands of heterosexual Chicanas. Carla Trujillo has railed against "the controlling forces of compulsory heterosexuality." At the 1989 MALCS workshop on feminism, when Chicana lesbians spoke out against the stifling heterosexism in the audience, members of the workshop seemed more interested in making men more sensitive than in examining their own attitudes and values. "Feminism is not about men washing dishes." Emma Pérez reveals the need for "safe" spaces. "So call me a separatist, but to me this is not about separatism. It is about survival."[93]

Since the days of the Chicano Student Movement, Chicana feminists have experienced fissures along the fault lines of sexuality, class, region, and acculturation. In *To Split a Human*, Carmen Tafolla has warned compañeras against competing for coveted spots in the hierarchy of oppression:

> Don't play. "Will the Real Chicana Stand Up?" Much as we have heard different groups compete for 'charter membership' in the

Most Oppressed Club . . . Mujer Sufrida ranks, and Double Minority Bingo, we must admit that membership dues must be con-
tinuously paid. . . . It is irrelevant to try to justify how "Chicana"
we are or to criticize others for being "Anglicized."[94]

Coming to terms with divisions in order to move forward as a collective remains both a dream and a challenge.

In 1982, Chicana undergraduates, graduate students, and
professors gathered at the campus of the University of California, Davis, to form Mujeres Activas En Letras Y Cambio Social
(MALCS). Called together by Adaljiza Sosa Riddell, chair of Chicano Studies at Davis, this small group of women, many from
Sacramento and the Bay Area, joined together to form a feminist
organization with a collective vision and responsibility to Chicanas
in the academy and the community.[95] A portion of the MALCS de-
claration follows:

The scarcity of Chicanas in institutions of higher education requires that we join together to identify our common problems, to
support each other, and to define collective solutions. Our purpose is to fight the race, class, and gender oppression we have experienced in the universities. Further we reject the separation of
academic scholarship and community involvement. Our research
strives to bridge the gap between intellectual work and active
commitment to our communities. . . . We declare our commitment to seek social, economic, and political change through our
work and collective action.[96]

Today MALCS is the most influential and largest Chicana/Latina
academic organization. It has convened annual conferences and
publishes a working paper series and most recently a journal, *Chicana Critical Issues*. MALCS has also served as an important feminist mentoring network for faculty, staff, graduate students, and
undergraduates.

"A leadership that empowers others" continues to the present.
Student activism has translated into community activism. In
southern California alone, many women exemplify the credo of giving back to your community. Judy Baca is an internationally re-
nowned muralist whose work brings Chicano history to life. María
Elena Durazo is the current president of the Hotel Workers' Union
in Los Angeles and Gloria Romero, an associate professor at Cal
State Los Angeles, spearheaded a campaign for citizen review of
the Los Angeles Sheriff's Department. She was recently elected as

a trustee for the Los Angeles Community College District. A blowout activist in 1968, Vicky Castro now serves on the Los Angeles County School Board and attorney Antonia Hernández heads the Mexican American Legal Defense Fund (MALDEF), arguably the most influential civil rights organization west of the Mississippi. Furthermore, Gloria Molina is the first Latina elected to the LA County Board of Supervisors.

But if the list of Chicana activists goes on across the West and Midwest, so do the battles for respect, recognition, and justice. In 1989, the UCLA Film and Television Department funded and screened a student film entitled "Animal Attraction" that depicted a "Mexicana/Chicana having sexual intercourse with a donkey." When Professor Gloria Romero and her friends voiced their objections at the screening, the film students booed them and one poured wine all over Romero. The campus police arrived—to arrest Romero. The southern California MALCS chapter staged a protest and in a letter to the chair of film and television, the *mujeres* demanded an investigation into the procedures through which such a film was funded; to match the funds given to the production with a grant to MALCS for its summer institute to be held at UCLA the following year; to support "a collaborate relationship" with the Chicano Research Center; and to admit more Chicanas into the film and television graduate program. MALCS also called for the creation of a Chicano Studies Department with a strong Chicana/Latina research component.[97] Addressing the department's Euro-American woman chair, they ended their letter as follows:

> Our voice is one which is frequently excluded through various discursive practices which honor a male/centered vision and policy, be it Anglo or Chicano. We hope you respond to our requests given that you refused to speak to our chapter representative yesterday. . . . We are the daughters of Guadalupe, Tenepal, and Adelita. We are the active women (professors, staff and students) for social change. We demand action now![98]

Such language resonates with the rhetoric of the Chicano Movement and summarizes the importance of reclaiming a distinctly Chicana heritage, of linking the past with the present.

The last three decades have witnessed the emergence of distinct Chicana feminist identities: collectively and individually. The road has not been easy. As scholar/activist Teresa Córdova has so eloquently stated, "Chicana feminists have struggled to find their voices—have struggled to be heard. Our struggle continues, but our silence is forever broken."[99]

Chicanas share a topography of multiple identities, and definitions of Chicana feminism remain contested. Instead of feminism per se, Ana Castillo calls for a *mestiza* consciousness or *Xicanisma*, an uncompromising commitment to social justice rooted in a woman-centered, indigenous past. Again with her global vision, Anzaldúa's construction of "the new *mestiza*" encompasses all women of color. In the preface to *Making Face, Making Soul*, she holds out a message of hope. "We are continuing in the direction of honoring others' ways, of sharing knowledge and personal power through writing (art) and activism, of injecting into our cultures new ways, feminist ways, mestiza ways."[100]

Bringing to their sociological study of Chicana feminism their experiences as student and community organizers, Beatríz Pesquera and Denise Segura divide Chicana feminists on college campuses during the 1980s into three distinct categories: Chicana Liberal Feminist, Cultural Nationalist Feminist, and Chicana Insurgent Feminist. "Chicana liberal feminism centers on women's desire to enhance the well-being of the Chicano community, with a special emphasis on improving the status of women." Cultural nationalists continue to see their lives as part of "the great family of Chicanos" or a feminism defined within traditional familial norms. Insurgent feminists take the intersections of race, class, gender, and power into consideration.[101] As one of their informants so cogently related:

> Chicana feminism means the struggle to obtain self-determina-tion for all Chicanas, in particular that Chicanas can choose their own life course without contending with the pressure of racism, sexism, and poverty. It means working to overcome oppression, institutional and individual. Chicana feminism is much more than the slogan: "the personal is political"; it represents a collective effort for dignity and respect.[102]

Chicana feminism also represents a commitment to community. "Feminism is necessary for liberation," Tejana historian Cynthia Orozco has so powerfully stated. "We must move beyond the barriers that the university seeks to maintain between a privileged sector and the mass of exploited and oppressed Mexicans. Sexism has no geographical barriers . . . nor should feminism stay in the college setting. Feminism belongs in the community." In the words of Gloria Anzaldúa, "Basta de gritar contra el viento—toda palabra es ruido si no está acompañada de acción" ["Enough shouting against the wind—all words are noise if not accompanied with action"].[103]

Figure 27 For over seventy years, Jesusita Torres (left) tilled a livelihood as a farm and nursery worker in southern California. Courtesy of Jesusita Torres.

Figure 28 Pecan shellers of San Antonio, Texas, 1938. The young woman standing in the polka-dot dress, Annie Pérez, is the mother-in-law of former mayor of San Antonio and HUD Secretary Henry Cisneros. Courtesy of Mary Alice Cisneros.

Figure 29 Women cannery workers at the California Sanitary Canning Company, 1936. Courtesy of Carmen Bernal Escobar.

Figure 30 UCAPAWA Local No. 3 negotiating committee, including Luisa Moreno (far left in plaid coat) and Carmen Bernal Escobar (third from left with hands around her child), 1943. Courtesy of Carmen Bernal Escobar.

Figure 31 The iconographies of a romantic Spanish past captured the imagination of affluent Anglos and Hispanos alike. The woman standing to the left in the nifty hat is New Mexico State Representative Concha Ortiz y Pino. Courtesy of New Mexico Cuarto Centennial Commission Collection, Photographer Harold D. Walter, The Center for Southwest Research, General Library, The University of New Mexico, Albuquerque, Neg. no. 000-048-0092.

Inside the mural:

VIVA LA

UNITED CANNERY
AGRICULTURAL PACKING &
ALLIED WORKE

EL CONGRESO DE PUEBLOS

LUISA MORENO

Figure 32 This section of Judy Baca's sweeping mural "The Great Wall of Los Angeles" honors the legacy of Luisa Moreno. When I gave Luisa this photograph, she seemed quite pleased with the artist's darkened rendition of her güera (light-hued) complexion. Courtesy of Judy Baca and Social and Public Art Resource Center.

Figure 33 A night on the town for civil rights advocate Josefina Fierro de Bright and her Hollywood screenwriter husband John Bright. Courtesy of Mytyl Glomboske.

Figure 34 Picketing and knitting: The real women of "Salt of the Earth." Empire Zinc Mining Strike, 1951. Courtesy of the Los Mineros Photograph Collection and Clint Jencks Papers, Chicano Research Collection, Department of Archives and Manuscripts, Arizona State University, Tempe.

Figure 35 Children at the line, Empire Zinc Mining Strike, 1951. Courtesy of the Los Mineros Photograph Collection and Clint Jencks Papers, Chicano Research Collection, Department of Archives and Manuscripts, Arizona State University, Tempe.

Figure 36 Defiant in the face of shotguns and billy clubs, Elvira Molano, co-chair of the union negotiating committee, was dubbed "the most arrested woman" during the Empire Zinc Mining Strike, 1951. Courtesy of the Los Mineros Photograph Collection and Clint Jencks Papers, Chicano Research Collection, Department of Archives and Manuscripts, Arizona State University, Tempe.

Figure 37 A child beside an outhouse, 1949. Such conditions spurred the growth of Mexican-American civil rights organizations after World War II. Courtesy of Lee (Russell) Photograph Collection, The Center for American History, The University of Texas at Austin, Neg. no. AD19-13198.

Figure 38 Children attending a segregated school in San Angelo, Texas, 1949. Courtesy of Lee (Russell) Photograph Collection, The Center for American History, The University of Texas at Austin, Neg. no. 14233-27.

Figure 39 Migrant elementary school student, 1963. Courtesy of the Denver Public Library, Western History Department.

Figure 40 Primarily a middle-class organization, the League of United Latin American Citizens (LULAC) has for over seventy years resisted discrimination in education, employment, and housing while holding fast to aspirations of the American Dream. Courtesy of the Mexican American Collection, Chicano Research Collection, Department of Archives and Manuscripts, Arizona State University, Tempe.

Figure 41 Migrant farm worker children, Colorado, ca. 1968. Courtesy of the
Denver Public Library, Western History Department.

Figure 42 Embracing cultural nationalism, Chicano students sought to reclaim a glorious Aztec past. Courtesy of the Denver Public Library, Western History Department.

Figure 43 Students represented the vanguard of the Chicano Movement: Mexican Independence Day March, Denver. Courtesy of the Denver Public Library, Western History Department.

Figure 44 Chicanos at the University of Colorado listen to a young lawyer named Frederico Peña. Courtesy of Juan Espinosa and the Colorado Historical Society, Denver.

Figure 45 Children of CASA line up to march. Historian Marisela Chávez, pictured in pigtails, stands in the second row. Courtesy of Department of Special Collections, Stanford University Libraries.

Bridging the academy and the community—Chicana feminist scholars/activists.

Figure 46 Gloria Romero, 1992. Courtesy of The Claremont Graduate School.

Figure 47 Mary Pardo, 1992. Courtesy of The Claremont Graduate School.

Figure 48 Returning to the community with your education was the informal credo of the Chicano Movement. Standing in front of the Stanford University chapel are the 1988 coordinators of the Latino Leadership Opportunity Program which encouraged undergraduates to consider careers in public affairs. Courtesy of Victor Becerra who stands at the far right, middle row.

Figure 49 Jesusita Torres in the yard of the home she bought for seventeen dollars during the Depression. Personal collection of the author.

6

Claiming Public Space

❖

DURING the second week of May 1972, Elsa Chávez, a twenty-six-year-old El Paso garment worker, left her post at the Paisano plant of Farah Manufacturing. Joining 4,000 Farah employees in El Paso, San Antonio, Victoria, Texas, and Las Cruces, New Mexico, Chávez became part of a twenty-two month strike for seniority rights, higher wages, pension issues, and union recognition. Management's manipulation of production quotas sparked the most indignation among Mexican women who formed the backbone of Farah's line personnel. Chávez's anger is easy to understand. Imagine that you have been hired to sew belts onto slacks. To get a raise, you must meet a quota of 3,000 belts per day, which translates into sewing six belts per minute. Forget your lunch hour, for if you fall too far behind in fulfilling your quota, you will be fired. As a "checker" of freshly sewn garments, Chávez felt fortunate in that she did not have to meet a quota; yet she understood the situation of the seamstresses with whom she worked. "Some of my fellow workers . . . were very nervous because they were always told, 'Hurry up' . . . 'You're going to be fired.' There was so much pressure that they started fighting for the bundles." Although Chávez perceived herself as earning good money and facing little personal harassment, she too walked out.[1]

Successful union organization depends, in large measure, on a sense of solidarity and community among workers. Effective political and community action requires the intertwining of individual subjectivities within collective goals. Claiming public space can involve fragile alliances and enduring symbols, rooted in material realities and ethereal visions. On a situational, grass-roots level, informal and formal voluntary organizations do serve as conduits for women's collective identity and empowerment. For the Mexican women whose voices foreground this study, the individual bonds formed at work, at church, or in the neighborhood reflect a mosaic of subjectivities, strategies, and goals but remain rooted in collective struggles for recognition and respect.

Gender and social justice—doesn't that equal feminist consciousness? As contemporary Mexican women know all too well, it depends on whose feminism and whose context. As one Farah striker bluntly stated, "I don't believe in burning your bra, but I do believe in having our rights."[2] Mediated by gender, race, culture, and class, activism transforms individual conceptions of self, changes that alter people's lives with subtlety or drama. Labor disputes raise the stakes in the precarious politics of work and family. Drawing on the insightful scholarship of Laurie Coyle, Gail Hershatter, and Emily Honig, on union pamphlets and newsletters, on local media coverage, and on oral history, an examination of the Farah strike provides a powerful case of community building through union organization.

When she joined her striking co-workers, Elsa Chávez had little inkling of the struggles that lay ahead. Farah Manufacturing was the largest private employer in El Paso; its chief executive officer Willie Farah had a reputation as a patriotic, civic-minded business leader; and labor activism found few friends in a conservative border city notorious as a "minimum wage town."[3] Indeed, the May 1972 walkout was not the culmination of an overnight organizing drive by representatives of the Amalgamated Clothing Workers (ACW), but the result of a protracted campaign begun in 1969. Events on the picket lines outside the El Paso plants quickly convinced the ACW that the strike could not be won within the city limits. Willie Farah responded with armed security guards walking with unmuzzled police dogs and he obtained a court order upholding an 1880 law stipulating that pickets must stand at least fifty feet apart. From 800 to 1,000 people, many of them women, were arrested, some during midnight raids at their residences. Instead of the usual $25 bail set for misdemeanor offenses, those arrested during the Farah strike were required to post a $400 bond.[4]

Within a few months, the 1880 picketing law was declared unconstitutional and Farah was ordered to "call off" the dogs and de-sist from interfering with peaceful picketing. Farah decided to take his case to the U.S. Supreme Court, but in August Justice Lewis Powell ruled against him by affirming the decisions of the lower court. At the same time the National Labor Relations Board charged Farah with unfair practices with regard to intimidation and harassment.[5]

But if Willie Farah found little solace outside his native El Paso, he found plenty of local support. Both El Paso dailies offered him a friendly hearing. And when El Paso's Catholic bishop Sidney Metzger openly supported the strikers, an area Protestant minister Paul Poling wrote a highly charged anti-union pamphlet endorsed by thirteen other clergy. Yet, despite being insulted at street corners where they picketed department stores, in their neighborhoods, and in letters to the editor, the Farah strikers continued their vigil.[6]

As one El Paso activist reflected, "We thought when we went out on strike that our only enemy was Farah . . . but we found out it was also the press, the police, the businessmen. . . . This strike was not just for union recognition."[7] The Farah strike distilled the racial and class cleavages within El Paso, cleavages evident in both the daily lives of the strikers and the opinion sections of local newspapers. Letters to the editor typically chastised the Farah strikers for their ingratitude, ignorance, and gullibility to outside agitators. Or as one retired El Paso retail saleswoman (F.T.T.) wrote,

> Mr. Farah did not invite one of the people who are working for him to come and work for him. They all asked for a job and should thank God that they got one. If they think they are such hot stuff and qualified for a better job then why don't they quietly fold their tents and leave? . . . The Farah family has worked hard for what they have and no-one has the moral right to harm them. I would like to give my boot—and you know where—to those picking [sic] in front of The Popular."[8]

Although their letters appeared less frequently than those of their opponents, strikers and their supporters responded in kind, and the editorial pages impart a sense of the polarity of opinions surrounding the "morality" of the Farah dispute. In a spirited letter critical of media coverage, Irma Camacho wrote, "The Farah struggle, since its conception was a moral fight for human dignity, since then through the use of your newspaper, the controlling 'powers that be' have stolen what objectivity there could exist and have subjectively used the facts to place economic value over human rights."[9]

Realizing that "the strike won't be won in El Paso," the Amalgamated Clothing Workers called for a national boycott of Farah suits and slacks.[10] Supported by AFL-CIO unions, campus activists, celebrities, and liberal politicians, Citizens Committees for Justice for Farah Workers sprung up in cities from the Pacific Coast to the Atlantic seaboard. In addition to holding fundraisers, these groups organized picket lines in front of department stores that carried Farah products and during "Don't Buy Farah Day" on December 11, 1972, an estimated 175,000 people, predominately AFL-CIO members, held rallies and parades across the country. United Farm Workers President Cesar Chávez visited the workers as did Sergeant Shriver, the Democratic vice presidential nominee of 1972. The Farah strikers also listened as U.S. Representative Edward Roybal from California spoke of his youth as a garment presser and ACW member. Support even crossed party lines with Nelson Rockefeller's public endorsement.[11]

The national boycott slowly began to have its desired effect. Sales declined from $150 million in 1972 to $126 million in 1974. As an article in *Texas Monthly* revealed, some retailers gladly took Farah slacks off their racks not in response to pickets but in retaliation for Willie Farah's "high-handed methods of doing business."[12] While the boycott undoubtedly contributed to bringing Farah to the bargaining table in March 1974, the dedication of the Mexicanas holding the line cannot be underestimated.

As weeks turned into months, the Farah picketers turned to one another for support, friendship, advice, and action. Critical of ACW support in El Paso, one group of women formed a rank-and-file committee within the union. According to Coyle, Hershatter, and Honig, the "members . . . shared a strong sense of themselves as workers and a desire to build a strong and democratic trade union." The women "put out their own leaflets, participated in marches and rallies, helped to found the Farah Distress Fund, and talked to other strikers about the need for a strong union." This caucus continued its work after the strike's successful settlement under the name *Unidad Para Siempre*.[13]

The strike, however, divided friends and families since a little less than one-half of the original workforce had walked out. Elsa Chávez recalled how this schism affected her personally.

> I had a fiancé there—we were going to get married, but he was from the inside, I was from the outside. So we broke up because

he didn't want anything to do with the strikers. After the strike,
he came back and (then) I told him "good-bye."[14]

Chávez continued, "But you wouldn't believe the number of di-
vorces caused by the strike. A lot of couples broke up either the
wife was inside and the husband was outside or the other way
around."[15]

As amply documented in "Women at Farah," tensions among
kin and friends took its toll and, as the strike wore on, financial
pressures mounted. Many families lost their homes, automobiles,
and other possessions. Although the union had an emergency
strike fund and distributed groceries and clothing, many Farah activists found themselves in severe economic straits. One woman
explained, "A lot of people lost their homes, cars—you name it,
they lost it." Such circumstances fueled marital tensions, but 85
percent of the strikers were women and they sought ways to balance picket and family duties.[16]

Children taking their place on the picket line occurred
throughout modern Chicano labor history, as early as the 1930s. As
an example, during the 1933 Los Angeles Dressmakers Strike, IL-
GWU representative Rose Pesotta organized 300 children in costume for an impromptu Halloween parade in front of the factories
where their mothers were picketing.[17] While perhaps mothers initially brought their children to the line because they had little
choice, the youngsters began to prove themselves useful in distributing leaflets outside stores. Adults were less likely to make abusive
comments toward a child.[18] In the words of Farah activist Julia
Aguilar, "Now, we just bring our children to our meetings, and we
bring them to the picket lines. Sometimes they ask, 'Are we going
to the picket line today, mommy?'" Aguilar continued, "It's kind of
hard with kids. But I'm willing to sacrifice myself and I think my
husband is beginning to understand."[19]

The settlement of the Farah Strike in March 1974 had, for
many women, come at great personal cost. Few activists would enjoy the benefits since many of the most vocal were fired after a few
months, ostensibly for failing to meet inflated production quotas;
union representatives blithely refused to generate any grievance
procedures to protect and retain these women.[20] Mexican women
have not fared well in their affiliation with mainstream labor
unions even though they have contributed much of the people
power, perseverance, and activism necessary for successful organi-

zation. As in the case of Farah, they typically have been denied any meaningful voice in the affairs of the local they had labored so valiantly to build.

Yet the Farah strikers had created community with one another and asserted their claims for social justice. As Coyle, Hershatter, and Honig wrote, "The Chicanas who comprise the majority of the strikers learned that they could speak and act on their own behalf as women and as workers, lessons they will not forget."[21] Elsa Chávez represents one of the women activists who have sought to merge personal and community empowerment. After she was fired, Chávez began to work at another clothing factory, but came to the realization that she wanted—and could achieve—a college education. I first met Ms. Chávez when she was a student in my Chicano history class at U.T. El Paso; two former strikers had enrolled in the class, a fact I discovered as I lectured on the Farah strike and noticed the two reentry women, both bilingual education majors, sitting in the front row winking and giggling to each other. "Oh, we're sorry, Dr. Ruiz, but we were *there*." A bit nonplussed, I turned the class over to them. Elsa Chávez dreamed of opening up "a school for slow learners" and had begun to organize a group of Mexican-American women in her education classes for this school.[22]

Labor struggles can also be centered around the involvement of the entire family. The United Farm Workers provides the most well-known example. Drawing on his experience with the Community Service Organization (CSO), Cesar Chávez in 1962 began to organize farm workers in the San Joaquin Valley. During the grape boycotts of the late 1960s and 1970s, Chávez and his United Farm Workers Organizing Committee (later the UFW) utilized tactics, such as the secondary boycott, national support committees, and identification with the Catholic Church. A charismatic leader preaching social justice and nonviolence, Chávez became the most prominent advocate for the rights of Mexicans in the United States. By 1973, the UFW "had contracts with 80% of the grape growers in the San Joaquin Valley" and "at its height the union had 100,000 members."[23]

UFW organizers, many of whom grew up as farm workers themselves, recognized that the family formed the unit of production in agriculture and consequently focused on the involvement of every family member. Referring to *campesinas*, Chávez related, "We can't be free ourselves if we can't free our women."[24] Signing union membership cards has always been a monumental decision for most farm workers because they risk not one job, but the live-lihood of the entire family. While the husband might be the first to sign the card, he often does so at the insistence of his wife. Former UFW legal department volunteer Graciela Martínez Moreno explained:

> At the beginning, women were more afraid of the union. But once they got the information about the benefits their children would receive, the women became good supporters. The biggest problem was getting through the initial fear, but if you got to the wife, the husband was sure to follow. The quiet, subtle pressure of the wife was very effective.[25]

But there is more to the story than this, for Mexican women have been well represented at the leadership and service levels of the United Farm Workers. Women, to a large extent, operate the service centers, health clinics, day nurseries, and legal departments.[26] Founding the first UFW Service Center in the Midwest in concert with her husband, María Elena Lucas offered a realistic portrayal of exhaustion. "And I worked such long hours, during the nights and on the weekends . . . but I just didn't know how to say no to people. I got very skinny. Sometimes I'd have thirteen or fourteen people waiting for me to do different things for them. It was just impossible." She continued, "Cesar had told me, 'It's not good to play Santa Claus to the people. It'll be neverending.' and I started getting to the point where I understood . . . I was burning out."[27]

Frustrated by the UFW's reluctance to organize migrant laborers in Illinois and exhausted from her job as a union social worker, Lucas became an organizer in 1985 with the Farm Labor Organizing Committee (FLOC) led by Baldemar Velásquez. Joined by four of her *compañeras*, including her own daughter, and their children (Gloria Chiquita had six kids), Lucas helped organize over 5,000 Midwestern farm workers and orchestrated a successful union election and contract. María Elena Lucas and her friend Gloria Chiquita both became vice-presidents in the FLOC, although Lucas expressed a feeling of powerlessness with regard to decision-making within the union board. She also recognized the difference in support systems between men and women organizers. Referring to men, she remarked, "They have the support of their wives and families, but most of us women have to work against our husbands and all of the services they expect."[28]

Fran Leeper Buss's oral history of María Elena Lucas, *Forged*

Under the Sun/Forjada bajo el sol provides the most nuanced portrait of a woman rank-and-file organizer. What emerges most vividly from Buss's skillful editing of Lucas's life story is the strength and comfort migrant women find in their friendships with one another. Lucas gives unvarnished testimony to the oppression and abuse women face in the fields and at times in the home and the union hall.[29]

In her pathbreaking dissertation, "Women in the United Farm Workers," Margaret Rose documents women's networks in the *campesino* centers, ranch committees, and boycott campaigns. Rose divides UFW women into two typologies—"nontraditional" (UFW vice-president Dolores Huerta) and "traditional" (Helen Chávez, Cesar's widow). Although Rose portrays Huerta as someone fitting her union nickname of "Adelita" (the symbolic *soldadera* of the Mexican Revolution), she notes how even the "nontraditional" Huerta relied on extended kin and women friends in the union (the "union family") to care for her eleven children during her frequent absences. Although criticized for putting la causa first, Dolores Huerta has had few regrets. As she informed Rose, "But now that I've seen how good they [my children] turned out, I don't feel so guilty."[30]

Dolores Huerta is a fearless fighter for social justice. In 1962, she taught school in Stockton, California while being a political activist with the CSO and a mother of six with a seventh on the way. "When I left my teaching job to go start organizing farm workers, a lot of people thought I had just gone completely bananas."[31] A tough, savvy negotiator, Huerta skillfully manipulated her positionality as a mother at the bargaining table.

> When I had my younger children and I as still negotiating, I would take nursing breaks . . . everybody would have to wait while the baby ate. Then I would come back to the table and start negotiating again . . .
>
> I think it made employers sensitive to the fact that when we're talking about benefits and the terms of a contract, we're talking about families and we're talking about children.[32]

Although they contributed in different ways, Dolores Huerta, more public, and Helen Chávez, more private, there was no separation of home and union. For Helen Chávez, the UFW became an extension of her familial responsibilities. She worked in the fields while her husband organized, took care of the children and household, and participated in the social service end of the union. In a rare interview, Helen Chávez offered a glimpse of daily family life: "I hadn't worked for a while, because at the time of the year you could only pick up a day's or a week's worth of work. . . . You just made a few hampers [of peas] and that was it."[33] She further recalled winning a supermarket contest:

> Everytime you went to Safeway, they would give you a little coupon. . . . Everytime we went to the store we saved these. . . . So when we got one of those little tags, I told the checker, "This is going to be my winner," and he laughed. I was just joking with him. I gave the tag to the kids . . . I think it was one of the girls who put saliva on it, and came in yelling, "Mom, Mom, you got the flag! You won!" . . . I rushed back to Safeway. I was really excited. I had won $100 and, oh boy, what a lot of food for the kids! After I got my check, I told Cesar, "Look, we can get some things." And he said, "I'm sorry, but this going to our gas bill." He said he was about to lose his gasoline credit card because he owed $180. I was so disappointed, I sat and cried. I made so many plans for that $100![34]

As the union grew, Helen Chávez left the fields and became the manager of the credit union. Her integration of family, work, and activism exemplifies the "political familialism" described by sociologist Maxine Baca Zinn. Many women, like Helen Chávez, preferred to work behind the scenes and, as Margaret Rose remarked, "Their contribution remains vital, but largely unrecognized."[35]

As Barbara Kingsolver's *Holding the Line* demonstrates, distinctions between traditional/nontraditional, striker/supporter, and Mexican/Euro-American can become blurred. Her study of the 1983 Phelps Dodge copper strike in Arizona pivots on workers' identities as longtime residents of the mining towns of Clifton and Morenci. There existed a mingling of families—Euro-American, Mexican, and Euro-American/Mexican—whose genealogies were as part of the mines as the shafts. While not to minimize the historical legacies of ethnic tension, the strikers of 1983 perceived themselves more along the lines of a class-based mining community. Arizona copper miners had long history of labor activism and, after World War II, viable union representation. According to Kingsolver, the women "had grown up with the union, a tool as familiar to them as a can opener or a stove."[36] Yet she also argues that although these women had knowledge of and some direct experience in labor disputes and a few had worked in the mines themselves, this strike changed their consciousness. Although they faced tear gas, arrests, and grave financial hardships, these women blocked traffic, took charge of the picket lines, and organized mass demonstrations. They kept their vigil for months. One law enforcement officer disingenuously remarked, "If we could just get rid of those broads, we'd have it made." Furthermore, many had gone to work for the first time and recorded feeling confident and independent as the result of outside employment. "I think there are a lot of feminists around here," Jessie Tellez informed Kingsolver. "There are some strong women here who won't ever go back to the way things were."[37] Cleo Robeledo put it this way, "Before, I was just a housewife. Now I am a partner." In *Changing Woman*, historian Karen Anderson argues that these small-town Arizona women acted out of an "attachment to their community" and "used the managerial and interpersonal skills they had developed as homemakers in order to organize . . . and mediate." Reminiscent of the shift in consciousness among the miners' wives of *Salt of the Earth*, this strike stands as another example of the fusion of the private and public spheres for collective goals. However, as both Rudy Acuña and Karen Anderson have pointed out, the strike did not have a rosy ending. The NLRB ordered an election in which only the scabs could vote and the union was decertified.[38]

Wages, benefits, safer working conditions, seniority, and union recognition are not the only reasons women will go out on strike. Regarding women's labor activism, modes of consumption can be as important as modes of production. In 1973, Tejana pecan shellers employed by McCrea and Son in Yancy, Texas, went out on strike for equal pay for equal work and for more sanitary conditions, but also because they resented "being coerced into buying 'Avon' products from the employer's wife."[39] Activism among Mexican women workers takes many forms, and the contours of their individual and collective agency shift in response to work environment, familial roles, and personal subjectivities.

In analyzing the rich history of labor activism among Mexican women in the United States, the transformation of women's consciousness, whether explicitly "feminist" or not, must be problematized through the shifting interplay of gender, race, class, culture, generation, and region. It is easier to celebrate the ways in which Mexican women have exercised control over their work lives than to examine the costs involved. In *Women's Work and Chicano Families*, anthropologist Patricia Zavella reveals the intricate sets of negotiations, networks, and decisions made by Mexican cannery workers in the Santa Clara Valley as they strived to build a rank-and-file caucus within the Teamsters' Union. Constructing a lucid, engaging narrative, she brings out the patriarchal infrastructures as well as attitudes on three levels—on the shop floor, in the union hall, and at home.[40] Zavella recognizes that neither family, neighborhood, the ethnic/racial community, nor union membership guarantees a comfortable "community."

In *Sunbelt Working Mothers*, Zavella, with co-authors Louise Lamphere, Felipe Gonzales, and Peter Evans, continues this discussion of work, family, and unions, with an emphasis on women's multiple networks. Focusing on "Hispana" and Euro-American factory workers in Albuquerque, the authors accentuate the importance of class and social location in building networks. "Our approach to ethnic and racial difference," they write, "focuses on behavioral strategies in response to material conditions, rather than exclusively on a cultural construction of ethnic identity."[41] However, these authors, among others, note that in the 1980s and 1990s, the transformation of women's work networks into effective union representation seems more elusive than ever. In recent years, runaway shops, anti-labor campaigns, and high-priced union-busting consultants, participative management styles, police harassment, mechanization, unemployment, and even the NLRB have stymied labor activism.[42]

While labor organizing, in general, has waned, union campaigns among service workers in Los Angeles show a remarkable vitality in building communities of resistance. Both "Justice for Janitors" (affiliated with the Service Employees International Union) and the Hotel and Restaurant Employees Union, Local 11, demonstrate the power of grass-roots organizing among the most politically and economically vulnerable sectors of the labor force—custodians, housekeepers, and food service workers, many of whom are undocumented Latino and Latina immigrants. In Los Angeles, those without union representation earn from $4.25 to $5.35 per hour—wages low enough for citizens "to qualify for food stamps."[43] Local 11's president María Elena Durazo is a college-educated Chicana for whom the idea of "going back to your community with your education" (a predominant theme of the Chicano Student Movement) was never empty rhetoric. Referring to immigrants as "the future of LA," Durazo eloquently stated, "I hope people will see unions as a tool for change and I hope we in the unions can respond to the challenge."[44]

With drive and conviction, Durazo leads a union that cannot be ignored. According to scholar/activist Mike Davis, Local 11 rewrites the book when it comes to organizing. "There will be no formal strike nor stationary picket line. . . . Across the city there will be leafletting, human billboards, flying pickets, delegations to city officials, and inevitably mass civil disobedience." Seeking a living wage, the members of Local 11 "speak of building not just a union, but a social movement like those of the 1930s and 1960s."[45]

As a labor historian, I will try to resist the temptation of privileging the workplace as *the* locus for claiming public space. Mexican women have relied on others as well, including, historically, the Roman Catholic Church. In *Hoyt Street*, her autobiography of growing up in Pacoima, Mary Helen Ponce portrays the local parish as the heart of her neighborhood in which time is recorded according to holidays and sacraments. In the barrios of the Southwest, Mexican women have been the stalwart volunteers for church fundraisers. At *jamaicas*, they sell tamales and *cascarones*, operate the cake walk, serve punch, organize the raffle, and help aspiring young anglers at the "go fish" booth. As feminist theologian Yolanda Tarango has argued, church activities were for many Latinas "the *only* arena in which they could legitimately, if indirectly, engage in developing themselves."[46]

Over the last twenty-five years, the Catholic Church, as both an institutional funding source (e.g., the Campaign for Human Development) and as community centers, began to support grassroots organizing campaigns among Mexican Americans. The most well known is the Alinsky-based Communities Organized for Public Service (COPS) in San Antonio, Texas. In 1973, with the support of local parishes, especially parish women, Ernie Cortés, Jr., began to organize neighbor by neighbor in San Antonio's Westside. He asked residents about their needs and concerns. This grassroots approach has permeated the infrastructure of COPS, with leadership emerging from local networks. Women's voluntary parish labor now became channeled for civic improvement and, indeed, several Tejanas have been elected president of the organization.[47]

Drainage problems and unpaved streets became their first order of business. Heavy rains made "peanut butter" of Westside roads; along with mud gushing into area homes and businesses, at times children died in "flash floods on their way to and from school."[48] Calling a public meeting with the city manager at the local high school in August 1974, COPS representatives caught him off guard with their numbers (over 500 people attended), their research, and their polite, yet firm queries. As former COPS president Beatrice Gallegos stated, "I sir'ed him to death." Through the use of demonstrations, political mobilization, research, and negotiation, COPS has significantly improved the material conditions of Westside and Southside neighborhoods. Focusing on municipal issues and boards, members of the twenty-five chapters of COPS ensure that developers, planners, school administrators, city officials, and Northside politicians do not ride roughshod over their communities. COPS has decisively influenced the distribution of Community Development Block Grants (CDBG) with "56 percent of the CDBG money allotted to San Antonio has gone to COPS-endorsed projects." They have also been active in local utility and environmental issues and opposed the funneling of over one million dollars of federal urban renewal funds into a suburban country club. COPS also engaged in voter registration drives and, while not endorsing candidates, its members closely monitored the positions taken by local politicians.[49]

The film *Adelante Mujeres* credits COPS with the 1981 election of Henry Cisneros as mayor of San Antonio, the first Tejano to hold the position since Juan Seguin in 1841. Yet the connection is not as clear it may first appear. COPS may have generated a level of political consciousness or civic engagement among Mexican Americans; Cisneros, who had significant crossover appeal among Euro-Americans, may have reaped the benefits of this heightened politicization.[50]

Unlike other community-based organizations in which leverage seems to rise and ebb, COPS is a respected grass-roots confederation with considerable municipal power. Scholars with divergent political perspectives have also recognized its importance as a model for community empowerment. Political scientist Peter Skerry praises COPS for the deftness with which the organization currently deals with development issues—relying on negotiation rather than confrontation. In defending Alinsky-style groups, such as COPS, from leftist criticism, historian/activist Rudy Acuña notes that without these groups "Many middle-class and poor Latinos would not be involved in social change programs." Or as political scientist Joseph Sekul put it, "COPS has taken giant steps toward raising the quality of life in older neighborhoods, some of which may now become places where people can stay if they choose, rather than leave because they must."[51]

Along with issues of family and neighborhood, COPS has cultivated the leadership of women. In the words of former COPS president Beatrice Cortez, "Women have community ties. We knew that to make things happen in the community, you have to talk to people. It was a matter of tapping our networks."[52] Unfortunately, Peter Skerry fails to appreciate the importance of women's civic labor as he impugns weakness in leadership to the organization's reliance on "housewives." Referring to what he considers "the authoritative role of organizers," he writes:

> Because organizers expend considerable time . . . working with them, leaders tend to find their involvement . . . quite stimulating. Unaccustomed to the sort of attention they receive, leaders typically experience marked personal growth. . . . But at the same time, these leaders . . . must be willing to put up with the organizers' demanding, sometimes harsh treatment. . . . For those who have a lot to learn, the bargain may seem a reasonable one. But for those with broader horizons and opportunities, it may not. As a result, the leaders . . . have been, with few exceptions, working-and lower-middle-class Mexican-American housewives with limited career prospects. These organizations have a much tougher time attracting college-educated Mexican Americans, especially well-educated men.[53]

Such condescension hardly merits elaboration. Skerry misses the significance of Mexican women's histories of community responsibility, education, and action. Or as Ernie Cortés simply stated, "COPS is like a university where people come to learn about public policy, public discourse, and public life."[54]

Feminist historian Sara Evans contextualizes Tejana participation in COPS within U.S. women's history. Bridging the public and private spheres, Evans argues that "women created a new public terrain through voluntary associations that became areas where citizenship could take on continuing and vital meanings, personal problems could be translated into social concerns, and democratic experiments could flourish."[55] While I agree with this supposition, I would also hasten to add that Mexican women's civic labor is neither recent in nature nor emulative in content. Examples of such activism can be located throughout the history of Mexican women in the United States. *Mutualistas*, like La Asociación Hispano-Americana; parish organizations, such as Hijas de María; middle-class auxiliaries, like the "ladies" of LULAC; and labor unions, such as UCAPAWA, provide strong evidence of women claiming public space for their community, their kin, and themselves.

With COPS as the model, similar organizations have emerged

throughout the Southwest. Although beyond the scope of this study, Los Angeles alone has at least three recognized and vital Alinsky-style community confederations: United Neighborhood Organization (UNO), South Central Organizing Committee (SC-OC), and East Valleys Organization (EVO).[56] Recently, members of SCOC tried to get Food 4 Less to build on a neighborhood lot owned by the discount grocery firm and the Community Redevelopment Agency. In the words of activist Orinio Ospinaldo,

> But having seen no progress at the Vermont site for many years, we were forced to take a dramatic step. . . . SCOC launched an "action" against Food 4 Less. We 65 adults and children went by bus to the Claremont [an upscale college community] office of Food 4 Less' chairman, Ron Berkel. His office was closed, so we distributed flyers outside. The leaflets compared the price of his home to the price of opening a new grocery store.[57]

Such direct action brought Berkel to meetings with the SCOC and Los Angeles Mayor Richard Riordan and, according to Ospinaldo, "Suddenly, things look promising." A single parent of three, Ospinaldo linked both family and community in the following remarks. "No matter where you live, that's your community and you have to fight to claim it. . . . But don't do it alone. You need the strength of people . . . [united] for a common good." He continued, "Second, try to involve your children. I've found an activity that is fulfilling and that acts as an example for my children."[58] Despite these articulated goals, the regional Catholic hierarchy has not always supported Latino and Latino/African-American community endeavors even when initiated by members of its own religious orders. *Calpulli* in San Bernardino provides such an example.

For over twenty years, Sister Rosa Marta Zarate and Father Patricio Guillen have pursued a vision of a dynamic *mesitizaje* of Latin American theology, Mesoamerican traditions, and community development projects. Forming Communidades Ecclesiales de Base (CEBs) in San Bernardino, Ontario, Riverside, San Diego, and the Imperial Valley, *calpulli* (the Aztec equivalent of neighborhood) fosters self-sufficient economic cooperatives. Unlike COPS, which works within the system, CEBs seek to build financially sustainable communities outside the arena of municipal politics. In some respects, *calpulli* represents an indigenous settlement house, offering classes in English, vocational education, and other commonly defined immigrant services. In addition, it closely resembles economic cooperatives found in Nicaragua, El Salvador, and other parts of Latin America. For example, *calpulli's* projects "include a travel agency, tax and legal counseling, book store, gardening and landscaping service, clothing manufacturing, and food service."[59] Scholars Gilbert Cadena and Lara Medina summarize Zarate and Guillen's efforts as follows:

> Today, they and a team of lay people successfully apply the tenets of liberation theology by creating a system of profit and non-profit cooperatives employing residents from the local community. Their goal is to create economically self-sufficient organizations that operate based on the principles of shared profit, shared responsibility, and shared power.[60]

Calpulli's successes have not gone unnoticed by church officials in Los Angeles. Several have looked askance at what they perceive as unorthodox community organizing. Indeed, Sister Rosa Marta Zarate received an ultimatum—return to Mexico or leave her order. She chose *calpulli* over the convent.

Zarate and Guillen also seek to make connections with southern California Native Americans and to educate project members in Mesoamerican history. Sister Rosa Marta Zarate reinterprets Aztec society from popular conceptions of feathered warriors and flamboyant sacrifices to understanding the economic cooperation that girded Aztec neighborhoods as well as an appreciation for their gendered spiritual values (e.g., recognizing Tonantzín, the earth mother). Preserving a historical memory also applies to contemporary activism. *Calpulli* has inaugurated an oral history project among its members.[61] El Plan de Acción de Calpulli encapsulates its mission as "an organization inspired by the cultures of our people, our history, our projects, and our destiny." Or as Lara Medina relates, "The underlying theme is living out their faith in a God who wants justice and humanity." She continues, "This faith motivates them to develop projects that will empower the personal and communal lives of la gente."[62]

Contemporary women's activism, however, does not necessarily revolve around the church. The work of sociologist Mary Pardo, for example, clearly delineates the networks of neighborhood organization among Mexican women in East Los Angeles. Founded in 1984, the Mothers of East Los Angeles (MELA) arose out of Resurrection Parish [no pun intended] to halt the construction of a prison in their community. This group of concerned women attracted 3,000 supporters as it staged demonstrations and rallies and engaged in political lobbying. Juana Gutiérrez summed up her involvement as follows, "I don't consider myself political. I'm just someone looking out for the community, for the youth . . . on the side of justice."[63]

State Assemblywoman Gloria Molina fervently supported their cause. She pointed out that the new prison would be built "within a four mile radius" of four correctional facilities and "within two miles" of twenty-six schools. Believing that enough is enough, Molina asked the rhetorical question:

> Do you think this could happen to Woodland Hills or Torrance? LA is supposed to have a prison, consequently, our community must bear the burden because we don't have the political strength to oppose it.[64]

However, the Mothers of East Los Angeles would forge that political strength.

While Peter Skerry characterized the women as housewives "led by a parish priest," Mary Pardo delineates Mexican women's organizing strategies that evolved independently of the Catholic Church. In her dissertation "Identity and Resistance: Mexican American Women and Grassroots Activism in Two Los Angeles Communities," Pardo offers compelling portraits of women as neighborhood activists, women who contextualize their civic labor as an extension of familial responsibilities Although considered as "political novices," the Mothers of East Los Angeles took on Governor George Deukmejian and the Department of Corrections and won. The prison was never built.[65]

From almost its inception, the Mothers of East LA have dedicated themselves to environmental issues. Their activities have ranged from leading the fight against a proposed incinerator to distributing free toilets to neighborhood residents. MELA has also raised money for scholarships and organized graffiti clean-up teams. The fusion of family and community resonates in the voices of these women. "The mother is the soul of the family; but the child is the heartbeat," Aurora Castillo, one of MELA's founders, explained. "We must fight to keep the heart of our community beating. Not just for our children, but for everyone's children."[66] Like Dolores Huerta and the women of the United Farm Workers, the Mothers of East LA have drawn on familial motifs for community and personal empowerment.

Mexican women's community activism is not limited to city streets. In his photojournal, *Organizing for Our Lives: New Voices from Rural Communities,* Richard Street poignantly documents the struggles of Mexican and Southeast Asian farm workers against toxic waste, pesticides, labor abuses, and discrimination in housing and education. Highlighting the activism among women, Street profiles several grass-roots associations represented by California Rural Legal Assistance. In chronicling organized protests against the building of an incinerator by Chem Waste in Kettleman City, Street photographs a young Mexican girl dressed in her frilly Sunday best. The Mexicanita is holding a large sign featuring Bart Simpson with the balloon caption, "DON'T HAVE AN INCINERATOR, MAN!"[67] This appropriation of an icon of U.S. popular culture represents a bifurcation of consciousness where the boundaries blur to the point that cultural codes converge in this subversion of the image.

In 1988, Mujeres Mexicanas, a campesina organization, was formed in the Coachella Valley. This group has participated in voter registration drives and electoral politics. Richard Street credits its members for the election of three Chicano city council candidates as well as the initiation of AIDS education in the valley. "They provided pamphlets, condoms, and bleach to disinfect needles. No local government or health agency in Coachella Valley was attempting anything like it."[68]

Many of the *mujeres* also belong to the United Farm Workers, in which Millie Treviño-Sauceda has been a rank-and-file organizer. In explaining the mission of Mujeres Mexicanas, Treviño-Sauceda revealed:

> Since the beginning we all agreed that our role was to promote the socio-political and psychological empowerment of *campesinas*. We also agreed that professional women—the ones with college educations—could only be advisors, not active members, because professionals tend to take over the leadership of the group. We wanted *campesinas* to be in control.[69]

The testimonies of the campesinas give witness to the power of women's collective action. In the words of María "Cuca" Carmona, "We have found our place within our community and even within our homes."[70]

Sustaining community space can be as important as finding it. For some areas, economic survival is resistance. In northern New Mexico, former SNCC volunteer and Chicano Movement activist María Varela has helped create and foster viable economic cooperatives among impoverished Hispano farmers, shepherds, and weavers. Los Ganados del Valle, which was founded in 1983, "operates on $150,000 annual budget and has 50 families as members." The cooperatives market yarns, quilts, clothing, and rugs; in 1990 its Tierra Wools subsidiary reached an annual sales of $250,000. Los Ganados del Valle has also organized around local environmental issues with regard to grazing rights. A recipient of a MacArthur Genius Award, Varela astutely contends, "I learned . . . that it is not enough to pray over an injustice or protest it or research it to death, but that you have to take concrete action to solve it."[71]

"Concrete action" resonates in the voices presented throughout this book. In examining women's activism, I am struck by the threads of continuity, the intertwining of community, family, and self. For some women, their involvement remains couched in familial ideology while others articulate feelings of personal empowerment *or* contexualize their actions within a framework of community-based feminism. Whether or not they proclaim feminist identities, their actions privilege collective politics over personal politics. Claiming public space, furthermore, can sustain, not subordinate, women's personal needs. Struggles for social justice cannot be boiled down to a dialectic of accommodation and resistance, but should be placed within the centrifuge of negotiation, subversion, and consciousness. Building community is both a legacy and a responsibility. As a storyteller, listener, recorder, and amateur theorist, I am reminded of a passage in Eudora Welty's *Becoming a Writer*:

> Each of us is moving, changing, and with respect to others. As we discover, we remember; remembering, we discover; and most intensely do we experience this when our separate journeys converge.[72]

Feminist theorist Chela Sandoval has adroitly distilled "the differential mode of oppositional consciousness" that underlies "concrete action." In her words: "The differential mode of oppositional consciousness depends upon the ability to read the current situation of power and of self-consciously choosing and adopting the ideological form best suited to push against its configurations."[73] In reflecting on her positionality in the hegemonic racial and economic structures of El Paso, Farah striker Estela Gómez addressed her grievances in a courageous letter to the editor:

> A lot of people in the El Paso community ask quite often, with all of these good benefits Willie Farah provides at his factory, why

did these people walk on strike? . . . These benefits were only there for the good of the company, not for the worker. . . . All these benefits put together could never make up for the only thing we are now struggling for and that is human dignity.

What good was the vaccuum [sic] cleaner he gave us for Christmas, when a lot of us didn't even earn enough to afford a carpet. . . . And the turkey for Thanksgiving—was it to make up for the time your supervisor made you cry because he wanted more production from you, as if you were a machine and not a human being? . . .

Be grateful to Farah they say, for all this man has done for you. I say Farah should be grateful to us, the Mexican-American, who from our sweat have [sic] worked hard to make the pants that have built his empire.[74]

Mexicana/Mexican American/Chicana activists, with determination, creativity, acumen, and dignity, have strived to exercise some control over their lives in relation to material realities and individual subjectivities as forged within both the spatial and affinitive bonds of community. Their courage comes forth out of the shadows.

Epilogue

IN 1995, as a registered California voter, I received a survey from the California group, American Immigration Control, a program of a nonprofit organization called "We The People," that asked respondents to name the social ills associated with immigration. Boxes to be checked included the following: terrorism, welfare fraud, taxpayer burdens, AIDS, drugs, riots, and bilingual education. The passage of Proposition 187 which is aimed solely at the immigrant, not the employer, heightens the possibilities for abuse by unscrupulous contractors and managers. Immigrant fears are certainly warranted. In a world with an enforced Proposition 187, undocumented mothers could be sentenced to five years in prison for sending their children to public school.[1]

"But I have promises to keep/And miles to go before I sleep."[2] Taken from a popular poem by Robert Frost, these two lines could serve as an anthem for Mexican women in the Southwest, women who, as a social and economic group, still find themselves at or near the bottom rung of the socioeconomic ladder. How large is this population? According to the Bureau of the Census, over twenty-five million Latinos (Puerto Rico included) live in the United States and Mexicans constitute over 61 percent of this growing population.[3]

By 1993, 51 percent of Mexican women in the United States held jobs outside the home.[4] Drawn primarily from census materials, the tables located in the Appendix illustrate quite starkly the disparities that exist in individual earnings, education, and household incomes. In Texas and California, where the majority of Mexican women live, they were at the *bottom* of the median income scale. Proportionally, Mexican women earn from forty-six cents (Texas) to fifty-five cents (New Mexico) for every dollar earned by Euro-American men. In general, Euro-American, Asian, African-American, and American Indian women have attained higher levels of education than either Mexican women or men. Only 5 percent of Mexican women in the labor force were college graduates, compared to 21 percent of their white, 31 percent of their Asian, and 13 percent of their African-American peers. Furthermore, as Table 5 illuminates, a "minority gap" in California exists where, in almost every educational category, Latinos earn less than other groups and Euro-Americans earn substantially more than comparably educated people of color.

In almost every instance, not surprisingly, female-headed families were financially disadvantaged in comparison with other families. Equally important, the standard of living for all households headed by women with children under six was, in general, well below the poverty line. According to the 1990 census, over 23 percent of Mexican families "were well below the poverty level." Twenty-eight percent of Mexican women, 19 percent of children, and 25 percent of the elderly live in poverty.[5]

Mexican women are more than twice as likely to be employed as blue-collar workers than their Anglo and African-American peers.[6] Mexican women with blue-collar jobs face the triple obstacles of gender, nationality, and class. For recent immigrants, citizenship poses an additional barrier. They are overworked and underpaid, often struggling to support themselves and their families on less than minimum wage. Cannery, garment, and other forms of factory work mean putting up with production speed-ups, sexual harassment, hazardous conditions, and substandard pay for fear that the plants will downsize or mechanize or pack up shop entirely and move "offshore" to Mexico, Costa Rica, or Honduras. As Patricia Zavella has pointed out, only automobile and aerospace industries outpaced food processing in terms of plant closures in California. "The number of canneries in the Santa Clara Valley dropped from a high of fifty-eight in 1930 to eleven in 1982." Some canners relocated to the San Joaquin Valley where they built high-tech, state-of-the-art facilities or moved them south of the border.[7]

As indicated previously, grass-roots organizing is far from dormant. One of the most compelling examples is based in San Antonio, Texas. Although Levi-Strauss recorded sales of $3.6 billion and profits exceeding $270 million, the company closed its South Zarazamora plant (the home of Dockers) in the process of relocating to Costa Rica. Eleven hundred workers (the majority Latina) lost their jobs. Launching a national boycott of Levi products, a rank-and-file caucus emerged. Reminiscent of Unidad Siempre during the Farah strike of the early 1970s, Fuerza Unida initiated a food bank, staged hunger strikes, and filed lawsuits in an attempt to extend pension benefits and severance pay. Betita Martínez has vividly recorded the voices of Fuerza Unida in her coverage for Z *Magazine.* "In the end, we saw that they treated the machines better than us." "'No tenemos hambre de comida, tenemos hambre de justicia'—we are hungry for justice, not for food." Although the Levi boycott has had neither the visibility nor success of the "Don't Buy Farah" campaign, Fuerza Unida represents the dignity of the human spirit in the face of seemingly unsurmountable realities associated with deindustrialization and offshore production.[8]

To many Americans, sweatshops are factories that exist in other nations, not our own. Homegrown garment sweatshops, however, did not disappear with Progressive reforms or New Deal legislation. Sociologist María Soldatenko has poignantly documented the exploitation of undocumented people in Los Angeles' garment shops. Managers and subcontractors intimidate workers from complaining about back pay or contract fraud by threatening to report them to the INS.[9] Pay scales, especially for homeworkers under the subcontracting system, seem analogous to the going rate during the Great Depression. One *costurera* was paid $3.75 for an intricately sewn cocktail dress. A 1994 survey of sixty-nine California garment plants indicated that 93 percent violated health and safety standards and over 50 percent violated minimum wage and overtime laws. Reminiscent of the Triangle Shirtwaist Factory in New York City at the turn of the century, some managers had locked or barricaded the exits and in two shops children "as young as 13 [were] working nine hour days."[10] Labor activist Angela Barcena aptly explains the economic vulnerability of Mexican blue-collar workers:

La mayoría de las mujeres en las fabricas, la mayoría necesitan
el trabajo para ayudar su esposo a mantener la familia o son las je-

fas de familia. Son las que mantienen su familia, son sola, son divorciadas. Así es que . . . la mujer han sido muy suspectible a la explotación-la mujer Mexicana.[11]

Yet there is hope in the faces of a new generation of activists, women like María Elena Durazo, Millie Treviño-Sauceda, and Yanira Merino. A thirty-one-year-old Salvadoran immigrant, Merino has shown remarkable courage and talent as a union organizer representing Laborers' International, "one of the nation's largest unions." In less than a year, she has gone from packing shrimp in downtown Los Angeles to unionizing poultry workers in North Carolina. A survivor of abduction, torture, and rape by unknown assailants in Los Angeles, Merino remains undaunted. Reflecting on her methods of organizing, she stated simply: "I don't think you can be an organizer behind a desk. You have to be out there relating . . . to the people so that you will learn. It's not having a theoretical understanding of what they're going through, but it's *knowing* what they're going through."[12] Yanira Merino also explains her work within the familiar intertwining of family and community. "I'll always see whatever work I do as an extension of my family." Working with migrant women, Millie Treviño-Sauceda emphasizes women's empowerment as individuals, especially in dealing with domestic violence. In her words: "Women . . . have rights, but these rights have no meaning or force unless we exercise them."[13]

Claiming public space can also involve claiming cultural space. A layering of generations exist among Mexicans in the United States from seventh-generation New Mexicans to immigrants. This layering provides a vibrant cultural dynamic. Artists Amalia Mesa Bains, Judy Baca, Yolanda López, and Carmen Lomas Garza and writers Sandra Cisneros, Denise Chávez, Pat Mora, and Cherríe Moraga (to name a few) articulate the multiple identities inhabiting the borderlands of Chicano culture. Entertainers can also influence cultural production. The queen of Tejano music, Selena Quintanilla Pérez, known simply as Selena, pushed cultural boundaries by blending sweetness and sensuality within a mestiza performance genre. She showed that a "nice girl" could be a sexy one as well. Only after her tragic death did a larger American public "discover" her music and her mystique. Although in less visible ways than a public stage, Chicana writers and artists, as chroniclers of consciousness, mark the daily rhythms of life. Poet/historian Naomi Quiñonez gives a sense of Los Angeles commutes in an excerpt from "The Confession."

> I was lost at rush hour
> for two of those three days
> and nirvana was a freeway offramp
> at 6th and Olympic
> where I found shelter
> at a 7-11, the urban temple of
> immediate gratification.[14]

Reminiscent of Luisa Moreno's "Caravan of Sorrow" speech,[15] over 1,500 artists, scholars, writers, and activists wrote an open letter condemning the beatings of undocumented immigrants by Riverside County deputies in April 1996. A passage follows:

> Just what crime did the beating victims, and the nineteen immigrants with them at the time, commit? They crossed the border between the U.S. and Mexico looking for work. They are among the immigrants who labor daily in the picking fields of California, Washington, and Pennsylvania. They clean hotel rooms in Dallas, toil in the canneries in Alaska, sell flowers on the street corners in New York City . . . the immigrants make life easier, prettier, and less expensive for those of us who have the luxury of calling ourselves "legals."[16]

Nativism has not gone unchallenged. The coalition against Proposition 187 and its implementation brought together high school and college students with labor and community groups, such as Justice for Janitors. On October 16, 1994, over 70,000 people marched in downtown Los Angeles to protest Proposition 187, "one of the largest mass protests in the city's history." Since 1900, Mexican women in the United States have sparked grass roots movements at work, church, neighborhood, and, within the last thirty years, on college campuses. Perhaps the motto of Fuerza Unida states it best. "La mujer luchando. El mundo transformando" ["Women in struggle transform the world"].[17]

Appendix

❖

Table 1 Occupational Distribution for Mexican Women in the Southwest, 1930–1990 (in percentges)

Occupational Group	1930	1950	1970	1990
White collar	15.4	32.4	43.9	54.5
Professional/technical	2.9	4.6	7.6	10.9
Managerial/proprietor	2.4	3.9	2.4	6.9
Clerical/sales	10.1	23.9	33.9	36.8
Blue collar	25.3	30.9	26.8	19.2
Skilled (Craft)	.6	1.4	2.2	3.7
Semiskilled (Operatives)	21.9	28.1	23.1	12.3
Unskilled (Laborers)	2.8	1.4	1.5	3.2
Service	38.4	27.8	26.2	23.9
Farm	20.7	6.5	3.1	2.4
Farm managers	1.0	.3	.1	.1
Farm workers	19.7	6.3	3.0	2.2

Source: Mario Barrera, *Race and Class in the Southwest: A Theory of Racial Inequality* (Notre Dame: University of Notre Dame Press, 1979), p. 131; U.S. Bureau of the Census, *1990 Census of Population, Social and Economic Characteristics: Colorado, Arizona, California, New Mexico, and Texas* (Washington, D. C.: G. P. O., 1993), Tables 50 and 124.

Table 2 Occupational Distribution for Mexican Women in the Southwest, by State, 1990 (in percentages)

Occupational Group	Colorado	Arizona	New Mexico	California	Texas
White collar	56.3	57.6	60.4	52.2	57.1
Professional/technical	11.9	11.0	13.6	9.4	12.8
Manager/proprietor	8.2	7.0	8.0	7.0	6.6
Clerical/sales	36.2	39.6	38.7	36.0	37.7
Blue collar	16.4	15.6	11.8	22.7	15.2
Skilled	3.6	3.7	2.4	4.2	3.1
Semiskilled	3.3	9.5	7.3	15.0	9.7
Unskilled	9.4	2.3	2.1	3.5	2.5
Service	26.6	25.3	26.8	21.6	26.8
Farm	.8	1.5	1.0	3.5	1.0
Farm managers	.05	.12	.07	.2	.08
Farm workers	.7	1.4	1.0	3.4	.9

Note: Farms include forestry and fishing.
Source: U.S. Bureau of the Census, *1990 Census of Population. Social and Economic Characteristics: Colorado, Arizona, California, New Mexico, and Texas* (Washington, D. C.: G. P. O., 1993), Tables 50 and 124.

Table 3 Years of School Completed by Selected Groups of Southwestern Workers, 25 Years and Older, 1990

Years of School Completed	Anglo Women	Asian Women	Black Women	American Indian Women	Mexican Women	Mexican Men
Elementary (8 yrs. or less)	8.3	16.2	9.1	16.1	37.5	37.1
High School (1–3 yrs.)	12.4	9.9	19.4	20.9	19.3	19.8
(4 yrs.)	27.3	18.8	25.7	26.6	21.1	19.1
College (1–3 yrs.)	31.2	24.3	32.4	28.4	17.1	17.5
(4 yrs. or more)	20.7	30.7	13.4	8.0	5.3	6.5

Source: U.S. Bureau of the Census, *1990 Census of Population. Social and Economic Characteristics: Colorado, Arizona, California, New Mexico, and Texas.* (Washington, D. C.: G. P. O., 1993), Tables 47 and 120.

Table 4 A Comparison of Median Incomes of Full-Time Workers 15 Years and over According to Gender, Ethnicity, and Race in the Southwest, 1990

	Median Incomes (in dollars)				
	Colorado	Arizona	New Mexico	California	Texas
Anglo Men	30,203	29,032	26,772	35,350	29,858
Mexican Men	20,416	19,594	19,153	19,699	17,255
Anglo Women	20,270	19,594	17,305	24,349	19,595
Asian Women	18,522	17,384	16,523	22,793	19,299
Black Women	18,785	17,551	16,108	22,996	16,682
Mexican Women	16,413	15,545	14,639	16,607	13,701
American Indian Women	16,992	15,197	14,274	20,714	17,191

	Median Incomes as Percentages of Total Anglo Male Median Income				
	Colorado	Arizona	New Mexico	California	Texas
Mexican Men	67.6	67.5	71.5	55.7	57.8
Anglo Women	67.1	67.5	64.6	68.9	65.6
Asian Woman	61.3	59.9	61.7	64.5	64.6
Black Women	62.2	60.5	60.2	65.1	55.9
Mexican Women	54.3	53.5	54.7	47.0	45.9
American Indian Women	56.3	52.3	53.3	58.6	57.6

Source: U.S. Bureau of the Census, *1990 Census of Population. Social and Economic Characteristics: Colorado, Arizona, California, New Mexico, and Texas* (Washington, D. C.: G. P. O., 1993), Tables 53 and 125.

Table 5 California Personal Income by Ethnicity and Education, 1989

Educational level	Anglo	Latino	Black	Asian
No high school diploma	$26,115	$16,487	$21,678	$18,517
High school diploma	$27,376	$21,121	$22,040	$21,608
Bachelor's degree	$44,426	$33,817	$34,290	$33,738
Master's degree	$52,787	$41,431	$42,254	$45,550
Doctorate	$59,348	$46,873	$54,205	$53,792
Professional degree	$77,877	$41,029	$61,015	$59,603

Source: Los Angeles Times, January 10, 1993.
Note: Professional degrees encompass law and medicine and Latinos can be of any race.

Table 6 Median Income for Families in Five Southwestern States, 1990

	Colorado	Arizona	New Mexico	California	Texas
Anglo two-parent	$37,228	$34,735	$30,341	$43,980	$35,080
female headed	$19,217	$19,451	$14,773	$24,213	$18,354
Asian two-parent	$32,286	$34,899	$32,211	$42,964	$35,729
female headed	$16,654	$14,286	$11,875	$25,977	$17,133
Black two-parent	$25,625	$24,120	$22,075	$29,453	$20,613
female headed	$13,206	$12,161	$9,918	$16,780	$11,572
Mexican two-parent	$22,708	$22,000	$19,979	$27,411	$19,807
female headed	$9,127	$11,764	$9,512	$15,226	$7,027
American Indian two-parent	$23,517	$14,015	$15,367	$30,620	$27,218
female headed	$11,442	$8,037	$9,264	$14,702	$13,325

Female Headed Households with Children Under 6

	Colorado	Arizona	New Mexico	California	Texas
Anglo	$8,921	$10,858	$7,094	$11,077	$11,487
Asian	$8,120	$5,000	$5,000	$12,051	$15,479
Black	$5,772	$5,000	$5,000	$8,893	$6,121
Mexican	$5,000	$5,000	$5,000	$8,463	$5,000
American Indian	$5,948	$5,964	$5,671	$8,038	$5,355

Note: $5,000 is a bottom-line figure adopted by the Census Bureau to indicate income at or below this level. The actual median incomes may be lower.
Source: U.S. Bureau of the Census, *1990 Census of Population. Social and Economic Characteristics: Colorado, Arizona, California, New Mexico, and Texas* (Washington, D. C.: G. P. O., 1993), Tables 53 and 125.

"La Nueva Chicana"
Viola Correa

Hey,
She that lady protesting injustice,
Es mi Mamá.
The girl in the brown beret,
The one teaching the children,
She's my hermana.
Over there fasting with the migrants,
Es mi tía.
These are the women who worry,

Pray, iron
And cook chile y tortillas.
The lady with the forgiving eyes
And the gentle smile,
Listen to her shout.
She knows what hardship is all about
All about.
The establishment calls her a radical militant.
The newspapers read she is
A dangerous subversive
They label her name to condemn her.
By the FBI she's called
A big problem.
In Aztlán we call her
La Nueva Chicana.

Source: Teresa Córdova, "Roots and Resistance: The Emergent Writings of Twenty Years of Chicana Feminist Struggle," in *Handbook of Hispanic Cultures in the United States: Sociology,* ed. Félix Padilla (Houston: Arte Público Press, 1994), p. 182.

University Avenue
Pat Mora

We are the first
of our people to walk this path.
We move cautiously
unfamiliar with the sounds,
guides for those who folllow.
Our people prepared us
with gifts from the land
    fire
        herbs and song
    hierbabuena soothes us into morning
    rhythms hum in our blood
    abrazos linger round our bodies.
    cuentos whisper lesson *en español.*
We do not travel alone.
Our people burn deep within us

Source: "University Avenue" in *Borders* by Pat Mora (Houston: Arte Público Press, 1986), p.19

Notes

Introduction

1. Vicki L. Ruiz, *Cannery Women, Cannery Lives: Mexican Women, Unionization, and the California Food Processing Industry, 1937–1950* (Albuquerque: University of New Mexico Press, 1987), p. xiv; Cleofas M. Jaramillo, *Sombras del pasado/Shadows of the Past* (Santa Fe: Ancient City Press, 1941). For a contemporary literary analysis of Jaramillo's works, see Genaro M. Padilla, *My History Not Yours: The Formation of Mexican American Autobiography* (Madison: University of Wisconsin Press, 1993) and Tey Diana Rebolledo, *Women Singing in the Snow: A Cultural Analysis of Chicana Literature* (Tucson: University of Arizona Press, 1995).

2. Alicia Arrizón, "Monica Palacios: 'Latin Lezbo Comic,'" in *Crossroads*, no. 31 (May 1993): 25.

3. *Ibid.*; Alicia Gaspar de Alba, "Literary Wetback," in *Infinite Divisions: An Anthology of Chicana Literature*, eds. Tey Diana Rebolledo and Eliana S. Rivero (Tucson: University of Arizona Press, 1993), p. 291.

4. Sonia Saldívar-Hull, "Feminism on the Border: From Gender Politics to Geopolitics," in *Criticism in the Borderlands: Studies in Chicano Literature, Culture, and Ideology*, eds. Hector Calderón and José David Saldívar (Durham, N. C.: Duke University Press, 1991), p. 220.

5. Chandra Talpade Mohanty, "Introduction: Cartographies of Struggle Third World Women and the Politics of Feminism," in *Third World Women and the Politics of Feminism*, eds. Chandra Talpade Mohanty, Ann Russo, and Lourdes Torres (Bloomington: Indiana University Press, 1991), p. 11.

6. Ramón A. Gutiérrez, *When Jesus Came, The Corn Mothers Went Away: Marriage, Sexuality, and Power in New Mexico, 1500–1846* (Stanford: Stanford University Press, 1991); Albert L. Hurtado, *Indian Survival on the California*

Frontier (New Haven: Yale University Press, 1988); Antonia I. Castañeda, "Presidarias y Pobladoras: Spanish-Mexican Women in Frontier Monterey, Alta California, 1770–1821" (Ph.D. dissertation, Stanford University, 1990); Antonia I. Castañeda, "Sexual Violence in the Politics and Policies of Conquest: Amerindian Women and the Spanish Conquest of Alta California," in *Building with Our Hands: New Directions in Chicana Studies*, eds. Adela de la Torre and Beatríz Pesquera (Berkeley: University of California Press, 1993), pp. 15–33; Angelina Veyna, "It is my Last Wish That . . ." A Look at Colonial Nuevo Mexicanas Through Their Testaments," in *Building with Our Hands*, pp. 91–108; James F. Brooks, "'This evil extends especially to the feminine sex': Captivity and Identity in New Mexico, 1700–1847" in *Writing the Range: Race, Class, and Culture in the Women's West* (Norman: University of Oklahoma Press, 1997) pp. 97–121.

7. Deena J. González, "Spanish-Mexican Women on the Santa Fe Frontier: Patterns of Their Resistance and Accommodation, 1820–1880" (Ph.D. dissertation, University of California, Berkeley, 1985); Sarah Deutsch, *No Separate Refuge: Culture, Class, and Gender on an Anglo-Hispanic Frontier* (New York: Oxford University Press, 1987); Lisbeth Haas, *Conquests and Historical Identities in California, 1769–1936* (Berkeley: University of California Press, 1995). A revised version of González's dissertation entitled *Refusing the Favor: Spanish-Mexican Women of Santa Fé, 1820–1880* is forthcoming from Oxford University Press.

8. Ellen Carol DuBois and Vicki L. Ruiz, *Unequal Sisters: A Multicultural Reader in U.S. Women's History* (New York: Routledge, 1990), p. xiii.

9. William Blake, "Auguries of Innocence" in *Poems of William Blake*, ed. W. H. Stevenson (London: Longman Group Limited, 1971), p. 585. I would like to thank British historian David Cressy for bringing this poem to my attention.

10. *Los Angeles Times*, September 27, 1992. *Note:* The exhibit was organized by El Monte Council member Ernie Gutiérrez. Noticing that the museum had rarely incorporated local Mexican–American history, he personally solicited photographs from long-time community residents. Museum staff mounted them for display and later returned the photos to the families without making copies for the museum's permanent collections.

Chapter 1

1. Interview with Jesusita Torres, January 8, 1993, conducted by the author.

2. Marjorie Sánchez-Walker, "Woven Within My Grandmother's Braid: The Biography of a Mexican Immigrant Woman, 1898–1982" (M.A. thesis, Washington State University, 1993), pp. 24–31, 33, 35–37, 39–40. Sánchez-Walker notes that family stories differ as to why the marriage took place. Indeed, Petra had expressed a desire to become a nun. Another Rocha sister, Luisa, would later explain that Petra had made "a solemn deathbed pledge" to Guadalupe that she would take her dying sister's place (p. 29).

3. Sánchez-Walker, "Woven Within My Grandmother's Braid," pp. 41, 44; Torres interview.

4. For any reader interested in the Spanish Borderlands, the essential first text is Ramón Gutiérrez's *When Jesus Came, the Corn Mothers Went Away: Marriage, Sexuality, and Power in New Mexico, 1500–1846* (Stanford: Stanford University Press, 1991). *Note:* Although some settlers would claim "Spanish" blood, the majority of people were *mestizo* (Spanish/Indian) and many colonists were of

African descent. [Quintard Taylor, "Exploration and Early Settlement" (unpublished chapter draft courtesy of the author), pp. 7–8.]

5. Helen Lara Cea, "Notes on the Use of Parish Registers in the Reconstruction of Chicana History in California Prior to 1850," in *Between Borders: Essays on Mexicana/Chicana History*, ed. Adelaida R. Del Castillo (Los Angeles: Floricanto Press, 1990), pp. 140–42.

6. For further reading on indentured servitude in the Spanish Borderlands, see Albert L. Hurtado, *Indian Survival on the California Frontier* (New Haven: Yale University Press, 1988); Douglas Monroy, *Thrown Among Strangers: The Making of Mexican Culture in Frontier California* (Berkeley and Los Angeles: University of California Press, 1990); Antonia I. Castañeda's "Presidarias y Pobladoras: Spanish-Mexican Women in Frontier Monterey, Alta California, 1770–1821" (Ph.D. dissertation, Stanford University, 1990); James F. Brooks, "'This evil extends especially to the feminine sex': Captivity and Identity in New Mexico, 1700–1847" in *Writing the Range: Race, Class, and Culture in the Women's West* (Norman: University of Oklahoma Press, 1997, pp. 97–121); Miroslava Chávez, "Don't Kill Me Sister: Gender, Race and Power in the Murder Trial of Guadalupe Trujillo in Mexican Los Angeles" (paper presented at the Berkshire Conference on the History of Women, University of North Carolina, Chapel Hill, June 7–9, 1996).

7. Monroy, *Thrown Among Strangers*, p. 271. Hubert Howe Bancroft offered the following description of women's work:

> They had charge of the kitchen and of the sewing which was by no means a light task, for there was a great deal of embroidery . . . In ironing the hand was used instead of a flat iron . . . They also combed and braided every day the hair of their fathers, husbands, and brothers. Many of them made the bread, candles, and soap consumed by the family, and many took charge of sowing and harvesting the crops.

[Hubert Howe Bancroft, *California Pastoral*, Vol. 34 of *The Works of Hubert Howe Bancroft* (San Francisco: The History Company Publishers, 1888), p. 312].

8. Marylynn Salmon, "Equality or Submission? Feme Covert Status in Early Pennsylvania," in *Women of America. A History*, eds. Carol Ruth Berkin and Mary Beth Norton (Boston: Houghton Mifflin, 1979), pp. 93–100; Janet Lecompte, "The Independent Women of Hispanic New Mexico, 1821–1846," in *New Mexico Women*, pp. 78–80; Alfredo Mirandé and Evangelina Enríquez, *La Chicana: The Mexican-American Woman* (Chicago: University of Chicago Press, 1979), p. 67. *Note:* Salmon also maintained that a few options (e. g., prenuptial contracts) did exist that allowed married women to control their property and other financial interests.

9. Albert Camarillo's *Chicanos in a Changing Society: From Mexican Pueblos to American Barrios in Santa Barbara and Southern California, 1848–1930* (Cambridge: Harvard University Press, 1979) provides the most comprehensive treatment of the downward mobility among Mexicans in the Southwest after 1848.

10. For more on stereotypes, see Tomás Almaguer, *Racial Fault Lines: The Historical Origins of White Supremacy in California* (Berkeley: University of California Press, 1994); Cecil Robinson, *With Ears of Strangers* (Tucson: University of Arizona Press, 1963); Deena J. González, "La Tules of Image and Reality: Euro-American Attitudes and Legend Formation on a Spanish-Mexican Frontier," in *Building with Our Hands: New Directions in Chicana Studies*, eds. Adela de la

Torre and Beatríz Pesquera (Berkeley: University of California Press, 1993), pp. 75–90. *Note:* Almaguer asserts that racialization and representation went hand in hand. Ethnic divisions among Europeans and Euro-Americans were put aside as they "railed against racialized groups in an attempt to arrogate for themselves a set of material interests that they ultimately defined as being their due as a 'white population'" [Almaguer, *Racial Fault Lines,* p. 210.]

11. Richard Griswold del Castillo, *La Familia: The Mexican American Family in the Urban Southwest* (Notre Dame: University of Notre Dame Press, 1984), pp. 81, 86; Mary Bernard Aguirre, "Public Schools of Tucson in the 1870s," Aguirre Family Papers, Arizona Historical Society Library, Tucson, Arizona. As an aside, New Mexico's *La revista católica* in 1877 editorialized that women's suffrage would destroy the family.

12. For further reading on Mexican women and work during the late nineteenth century, see Deena J. González, "Spanish-Mexican Women on the Santa Fe Frontier: Patterns of Their Resistance and Accommodation, 1820–1880" (Ph.D. dissertation, University of California, Berkeley, 1985); Sarah Deutsch, *No Separate Refuge: Culture, Class, and Gender on an Anglo-Hispanic Frontier* (New York: Oxford University Press, 1987); and Albert Camarillo, *Chicanos in a Changing Society.*

13. Fabiola Cabeza de Baca, *We Fed Them Cactus* (Albuquerque: University of New Mexico Press, 1954), p. 60.

14. Precise population and immigration figures for Mexicans in the United States between 1900 to 1930 do not exist. Oscar Martínez estimates that 375,000 to 552,000 Mexicans lived in the Southwest in 1900. In *Mexicans in California,* a 1930 study commissioned by Governor C. C. Young, the number of Mexican-born individuals in the United States reached 103,393. This figure includes only those born in Mexico and not the individuals born in the Southwest. See also T. Wilson Longmore and Homer L. Hitt, "A Demographic Analysis of First and Second Generation Mexican Population of the United States: 1930," *Southwestern Social Science Quarterly* 24 (September 1943): 138–48. Contemporary scholars posit that over one million (and perhaps as many as two million) Mexicanos migrated to the United States between 1910 and 1930. Ricardo Romo and Albert Camarillo opt for the lower end; George Sánchez and David Gutiérrez rely on estimates of 1.5 million; and Devra Weber uses the two million figure. See Oscar Martínez, "On the Size of the Chicano Population: New Estimates, 1850–1900," *Aztlán* (Spring 1975): 56; California, Governor C. C. Young's Mexican Fact-Finding Committee, *Mexicans in California* (San Francisco: California State Printing Office, 1930; rpt. R and E Research Associates, 1970), p. 31; Ricardo Romo, *East Los Angeles: History of a Barrio* (Austin: University of Texas Press, 1983), pp. 42, 61; Camarillo, *Chicanos in a Changing Society,* pp. 200–201; George J. Sánchez, *Becoming Mexican American: Ethnicity, Culture, and Identity in Chicano Los Angeles, 1900–1945* (New York: Oxford University Press, 1993), p. 18; David Gutiérrez, *Walls and Mirrors: Mexican Americans, Mexican Immigrants, and the Politics of Ethnicity in the Southwest, 1910–1986* (Berkeley: University of California Press, 1995); Devra Weber, *Dark Sweat, White Gold: California Farm Workers, Cotton, and the New Deal* (Berkeley: University of California Press, 1994), p. 52.

15. Gutiérrez, *Walls and Mirrors,* p. 6. I see the layering of generations in the faces of my students who are first, second, third, fourth generations, and beyond. For some, their border journeys were lived experiences and not something they learned about from books and family stories.

16. Ruth Tuck, *Not with the Fist* (New York: Harcourt Brace, 1946), pp. 209–10.

17. Interview with Ray Buriel, December 21, 1994, conducted by the author.

18. Chain migration generally refers to groups of relatives moving in stages to a specific locale—one family sets out to a new area and other relations gradually follow. Circular migration implies that families move back and forth across the border, with migration as a continuous process rather than as a prelude to permanent settlement. For more discussion, see George Sánchez, *Becoming Mexican American*, pp. 41, 132–33. Emilio Zamora in his study of Mexican workers in Texas also finds evidence of substantial circular migration along the Mexico–Texas border, a pattern that significantly shaped their cultural and political consciousness or what Zamora has termed a "Mexicanist identity." See Emilio Zamora, *The World of the Mexican Worker in Texas* (College Station: Texas A&M Press, 1993), especially Chapter 8.

19. In her contemporary study of Mexican immigrant women, Pierrette Hondagneu-Sotelo astutely reminds immigration scholars to look beyond an abstract structural analysis of push/pull factors in delineating the myriad of individual motivations for migration. In her words: "Conspicuously absent from the macrostructural perspective is any sense of human agency or subjectivity. Rather than human beings, immigrants are portrayed as homogenous, nondifferentiated objects, responding mechanically and uniformly to the same set of structural forces." [Pierrette Hondagneu-Sotelo, *Gendered Transitions: Mexican Experience of Immigration* (Berkeley: University of California Press, 1994), p. 6.] I have endeavored in the discussion that follows to tell women's stories, to account for the diverse ways in which they approached migration and settlement, to reclaim their motivations, expectations, and lived experiences.

20. Lawrence A. Cardoso, *Mexican Emigration to the United States, 1897–1931: Socio-Economic Patterns* (Tucson: University of Arizona Press, 1980), pp. 1, 7, 9–11, 13; Devra Weber, *Dark Sweat*, pp. 50, 52 [quote is from p. 50]; Linda B. Hall and Don M. Coerver, *Revolution on the Border: The United States and Mexico, 1910–1920* (Albuquerque: University of New Mexico Press, 1988), pp. 11–13. Hall and Coerver note that in "1910, foreigners—mostly Americans—owned about one-seventh of the land surface of Mexico. Much of this land was located . . . along the U.S.–Mexican border" [p. 12]. For more information on the modernization of Mexico under the Diaz regime, see Michael Meyer and William Sherman, *The Course of Mexican History*, 5th ed. (New York: Oxford University Press, 1995), pp. 431–67.

21. Meyer and Sherman, *The Course of Mexican History*, pp. 498–502, 552–61; Sánchez, *Becoming Mexican American*, pp. 33–34; Cardoso, *Mexican Emigration*, pp. 38–44; Elizabeth Salas, *Soldaderas in the Mexican Military: Myth and History* (Austin: University of Texas Press, 1990), pp. 36–48. For further information on Mexican women's revolutionary roles, see also Shirlene Soto, *Emergence of the Modern Mexican Woman: Her Participation in Revolution and Struggle for Equality, 1910–1940* (Denver: Arden Press, 1990) and Anna Macias, *Against All Odds: The Feminist Movement in Mexico to 1940* (Westport, Conn.: Greenwood Press, 1982).

22. Interview with Lucia R., December 1981, conducted by Margarita C.

23. Meyer and Sherman, *Course of Mexican History*, pp. 587–89; Cardoso, *Mexican Emigration*, pp. 38–49; Francisco Balderrama and Raymond Rodríguez,

Decade of Betrayal (Albuquerque: University of New Mexico Press, 1995), pp. 12–13; Ricardo Romo, "Mexican Americans in the New West," in *The Twentieth Century West,* eds. Gerald D. Nash and Richard Etulain (Albuquerque: University of New Mexico Press, 1989), p. 128; Interview with Alma Araiza García, March 27, 1993, conducted by the author.

24. See Albert Camarillo, *Chicanos in a Changing Society*; Romo, *East Los Angeles*; and Mario Barrera, *Race and Class in the Southwest* (Notre Dame: University of Notre Dame Press, 1979).

25. Barerra, *Race and Class in the Southwest,* p. 131.

26. Romo, "Mexican Americans," p. 126; Cardoso, *Mexican Emigration,* pp. 20–23; United States Chamber of Commerce, Immigration Committee, *Mexican Immigration* (Washington, D.C.: U.S. Chamber of Commerce, 1930), pp. 3, 8, 27, 32, 38. [Quote is from p. 32.]

27. "Letter to Anthony Caminetti, Commissioner General of Immigration from Y. Bonillas, Mexican Ambassador to the United States," Reel 8, Records of the Immigration and Naturalization Service, *Part 2: Mexican Immigration, 1906–1930* (University Publications of America microfilm edition), p. 32; "Labor Contract No. 176, Utah-Idaho Sugar Company, March 14, 1918," Reel 8 of *INS, Mexican Immigration,* p. 41.

28. Camille Guerin-Gonzales, *Mexican Workers and American Dreams: Immigration, Repatriation, and California Farm Labor, 1900–1939* (New Brunswick, N. J.: Rutgers University Press, 1994), p. 45; Paul S. Taylor, *A Mexican-American Frontier: Nueces County, Texas* (Chapel Hill: University of North Carolina Press, 1934), p. 103; Jean Reynolds, "Mona Benitez Piña: An Oral History of a Chicana Farmworker in Phoenix," *Palo Verde,* 3:1 (Spring, 1995): 67.

29. Erasmo Gamboa, "Oregon's Hispanic Heritage," *Oregon Humanities* (Summer 1992): 5; Letter from District Headquarters, U.S. Commission of Immigration, Moscow, Idaho, to Commissioner General of Immigration, Reel 8 of *INS, Mexican Immigration,* p. 83; Chamber of Commerce, *Mexican Immigration,* pp. 9, 23.

30. Jeff Garcílazo, "Taylorism, Patriarchy and *Traqueros:* Mexican Railroad Workers in the United States, 1880–1930" (unpublished paper), pp. 2–3; Guerin-Gonzales, *Mexican Workers,* pp. 33–42. Guerin-Gonzales delineates how the U.S. prohibition against contracting labor in Mexico was selectively enforced. Border inspectors only targeted those agencies owned by Mexican Americans.

31. Interview with Ray Buriel, December 21, 1994, conducted by the author; interview with Eusebia Buriel, January 16, 1995, conducted by the author; Chamber of Commerce, *Mexican Immigration,* pp. 22, 40; Garcílazo, "Taylorism, Patriarchy" p. 2; Jeff Garcílazo, "*Traqueros:* Mexican Railroad Workers in the United States, 1871–1930" (Ph.D. dissertation, University of California, Santa Barbara, 1995), pp. 192–245.

32. Rudolfo Acuña, *Occupied America: A History of Chicanos,* 2nd edition (New York: Harper & Row, 1981), p. 89; Sarah Deutsch, *No Separate Refuge: Culture, Class, and Gender on an Anglo-Hispanic Frontier in the American Southwest, 1880–1940* (New York: Oxford University Press, 1987), pp. 32, 88, 94; Patricia Preciado Martin, *Songs My Mother Sang for Me: An Oral History of Mexican American Women* (Tucson: University of Arizona Press, 1992), pp. 199–201; interview with Erminia Ruiz, February 18, 1993, conducted by the author; Manuel Gamio, *Mexican Immigration to the United States* (Chicago: University of Chicago Press, 1930; rpt. Arno Press, 1969), p. 39.

33. Paul S. Taylor, *Mexican Labor in the United States,* Vol. II (Berkeley:

University of California Press, 1932; rpt. Arno Press, 1970), pp. 1–11; Barrera, *Race and Class*, pp. 86–87; Gamio, *Mexican Immigration*, pp. 24–25; panel presentation by Yolanda López as part of "Puro Corazón: A Symposium of Chicana Art" held at Pomona College, February 16, 1995; Balderrama and Rodríguez, *Decade of Betrayal*, pp. 6–7.

34. Louise Año Nuevo Kerr, "The Chicano Experience in Chicago, 1920–1970" (Ph.D. dissertation, University of Illinois, Chicago, 1976), p. 20; Gamio, *Mexican Immigration*, pp. 16–19; Barrera, *Race and Class*, p. 66.

35. Weber, *Dark Sweat*, pp. 58–59; Manuel Gamio, *The Life Story of the Mexican Immigrant* (Chicago: University of Chicago, 1931; rpt. Dover Publications, 1971); "Biographies and Case Histories," Manuel Gamio Field Notes, Bancroft Library, University of California, Berkeley, 1 box. [see especially "Juana Martínez (Folder I), "Leova López" and "Elisa Morales" (Folder II)]; Reels 8– of *INS, Mexican Immigration.*

36. Zamora, *The World of the Mexican Worker*, pp. 18–19; Guerin-Gonzales, *Mexican Workers*, pp. 38–42; Sánchez, *Becoming Mexican American*, pp. 56–58. *Note:* Families headed by men were also recruited by contractors, notably those representing agribusiness,

37. Torres interview.

38. *Ibid.*

39. John Reed, *Insurgent Mexico* (New York: D. Appleton and Co., 1914; rpt. International Publishers, 1969). Rebozo means shawl.

40. "Hearing, Case No. 1439," Board of Special Inquiry, Nogales, Arizona, July 3, 1917, Reel 8 of *INS, Mexican Immigration*, p. 749.

41. *Ibid.*, 749–51. *Note:* Although Alfonso declared his desire for U.S. citizenship, he was only escorting his sisters to El Paso and expressing no intention of establishing a residence there.

42. Letter to Inspector in Charge, Immigration Service, Nogales, Arizona from Acting Supervisor Inspector, El Paso, dated July 20, 1917, Reel 8, *INS, Mexican Immigration*, p. 744. A portion of the reprimand follows: "As you know, the Mexican dancers and singers employed in cheap variety shows are, for the most part, of an irresponsible, immoral class, and it seems to the writer in this case an investigation properly should have been conducted at El Paso to determine, if possible, the character of the relatives living here and their mode of living before favorable actions was taken upon the applications."

43. Mario T. García, *Desert Immigrants: The Mexicans of El Paso, 1880–1920* (New Haven: Yale University Press, 1980), pp. 2, 46–48. Sánchez, *Becoming Mexican American*, pp. 55–57, 59 [quote is from p. 56]; Balderrama and Rodríguez, *Decade of Betrayal*, p. 9. The head tax was $8.00 per person. Because of a perceived labor shortage by southwestern growers, the literacy test and other provisions of the 1917 Act were waived for agriculture-bound workers. [García, *Desert Immigrants*, p. 47.]

44. García, *Desert Immigrants*, p. 2; Paul S. Taylor, *Mexican Labor*, Vol. II, p. vi. The original text in Spanish is as follows:

> Pues me decían que aqui los dólars
> se pepenaban y de a montón
> que las muchachas y que los teatros
> y aquí todo era vacilón.

45. Manuel Gamio, *The Life Story*, pp. 159–62; Torres interview [quote is from p. 159].

46. *Ibid.*

47. Ruth Allen, "Mexican Peon Women in Texas," *Sociology and Social Research*, Vol. 16 (September–August 1931–1932): 131.

48. Vicki L. Ruiz, *Cannery Women, Cannery Lives: Mexican Women, Unionization, and the California Food Processing Industry, 1930–1950* (Albuquerque: University of New Mexico Press, 1987). For a richly detailed family and community ethnography, but one that perhaps overaccentuates the positive, see Robert R. Alvarez, Jr., *Familia: Migration and Adaptation in Baja and Alta California, 1800–1975* (Berkeley: University of California Press, 1987).

49. Torres interview.

50. Interview with Julia Luna Mount, November 17, 1983, conducted by the author.

51. Torres interview.

52. Gilbert González, *Labor and Community: Mexican Citrus Worker Villages in a Southern California County, 1900–1950* (Urbana: University of Illinois Press, 1994), especially Chapters 2 and 3, pp. 43–98; Zamora, *The World of the Mexican Worker*, p. 5; Weber, *Dark Sweat*, p. 60.

53. Rosalinda González, "Chicanas and Mexican Immigrant Families 1920–1940: Women's Subordination and Economic Exploitation," in *Decades of Discontent: The Women's Movement, 1920–1940*, eds. Lois Scharf and Joan M. Jensen (Westport, Conn.: Greenwood Press, 1983), pp. 63–64, 72–73. Gilbert González contends that *compadrazgo* (ties of godparenthood for men and women) created a "regional network" among Mexican citrus workers in Orange County. [González, *Labor and Community*, pp. 75–76.] Referring to women's participation in these networks, Devra Weber writes, Women "cooked for each other when they were sick, helped take care of each other's children, provided support among themselves, and facilitated mutual aid within the community." [Weber, *Dark Sweat*, p. 66.]

54. Irene Castañeda, "Personal Chronicle of Crystal City: Part II," in *Literatura Chicana*, eds. Antonia Castañeda, Tomás Ybarra-Frausto, and Joseph Sommers (Englewood Cliffs, N. J.: Prentice-Hall, 1972), p. 247.

55. González, *Labor and Community*, p. 57; Interview with Clemente Linares conducted by Lydia Linares Peake, June 11, 1992; Mary Romero and Eric Margolis, "Tending the Beets: Campesinas and the Great Western Sugar Company," *Revista Mujeres*, 2 (Junio, 1985): 21.

56. "Chicano!: History of the Mexican American Civil Rights Movement," proposal submitted to the National Endowment for the Humanities by the National Latino Communications Center (August 1994), Appendix B, p. 4.

57. Interview with María Arredondo, March 19, 1986, conducted by Carolyn Arredondo.

58. Torres interview; Linares interview.

59. Isabel Flores, "I Remember," *El Grito*, Vol. 7 (September 1973): 80.

60. "Copies of Newspaper Clippings on the Deplorable Conditions Under Which Mexicans Are Forced to Live and Work in the State of Michigan," Box 14, File 2, Carey McWilliams Collection [the older material, call number 1243], Special Collections, University of California, Los Angeles, Los Angeles, California; "Migratory Workers of Southwestern Michigan," Prepared by John DeWilde, Marguerite Dewan, Ben Graham, and Bernard Litwin, August 5, 1940, Michigan Works Projects Administration, Box 14, File 2, McWilliams Collection. [Quote taken from p. 9.]

61. "Copies of Newspaper clippings," McWilliams Collection, p. 2.

62. Ray Buriel interview; interview with Josefina Fierro de Bright, August 7, 1977, conducted by Albert Camarillo; Ray Buriel, "The Cultural Adaptation of Hispanic Immigrants in the Citrus Belt, 1920–Present" (unpublished paper courtesy of the author); I. Castañeda, "Personal Chronicle," *loc. cit.*; Vicki L. Ruiz, "By the Day or the Week: Mexicana Domestic Workers in El Paso," in *Women on the U.S.–Mexico Border: Responses to Change* (Winchester, Mass.: Allen and Unwin, 1987), pp. 61–76. Irene Castañeda remembered her mother's laundry routine. "When she was washing clothes, she would sit us down beside her and she taught us to read Spanish."

63. Paul S. Taylor, *Mexican Labor in the United States,* Vol. I (Berkeley: University of California Press, 1930, rpt. Arno Press, 1970), pp. 356–57.

64. Torres interview.

65. Evelyn Nakano Glenn, "From Servitude to Service Work: Historical Continuities in the Racial Division of Paid Reproductive Labor," *Unequal Sisters,* 2nd ed., eds. Vicki L. Ruiz and Ellen Carol DuBois (New York: Routledge, 1994), p. 425.

66. Deutsch, *No Separate Refuge,* pp. 51–53, 131–32; González, *Labor and Community,* p. 66; Linares interview.

67. Romero and Margolis, "Tending the Beets," p. 26.

68. John Steinbeck, "Their Blood Is Strong," Simon J. Lubin Society pamphlet (1938), pp. 2–4, 21–22, 36–38; Arredondo interview.

69. Devra Anne Weber, "*Raiz Fuerte:* Oral History and Mexicana Farmworkers," *Unequal Sisters,* 2nd ed., p. 399.

70. "Labor Contract No. 176, Utah-Idaho Sugar Company, March 14, 1918," Reel 8 of *INS, Mexican Immigration,* p. 41; Report of Fred A. Caine, Utah-Idaho Sugar Company labor agent, January 25, 1918, Reel 8 of *INS, Mexican Immigration,* pp. 108–10; "Letter to Anthony Caminetti, Commissioner General of Immigration from Wilson J. McConnell, District Head, U.S. Commission of Immigration, Moscow, Idaho, Reel 8 of *INS, Mexican Immigration,* pp. 83–86. [Quotes from pp. 86 and 83, respectively.]

71. Letter to Charles L. Andrews, Inspector-in-charge, U.S. Immigration Service, Helena Montana, dated January 4, 1918, from Utah-Idaho Sugar Company, Reel 8 of *INS, Mexican Immigration,* pp. 88–89. Immigration officials did indeed search for these families in Idaho and Nevada. If apprehended, they would have two choices: return to the company or face deportation. [Letter to Immigration Inspector, Moscow, Idaho, dated January 27, 1919, from Office of Supervising Inspector, El Paso, Reel 8 of *INS, Mexican Immigration,* pp. 99–100.]

72. González, *Labor and Community,* pp.75–98; Buriel, "The Cultural Adaptation of Hispanic Immigrants." [Quote is from González, *Labor and Community,* p. 75.]

73. Eusebia Buriel interview; González, *Labor and Community,* p. 76.

74. Ray Buriel interview.

75. Garcílazo, "*Traqueros,*" p. 192.

76. *Ibid.,* pp. 208–12, 217–18. [Quote is from p. 212.]

77. *Ibid.,* pp. 222–23.

78. *Ibid.,* pp. 210–11; John Culhane, "Hero Street, USA," *Reader's Digest* (May 1985): 82. Gracias a Jesus Treviño for providing me with this article.

79. Ray Buriel interview; Eusebia Buriel interview.

80. "Un cuento—una vida," unpublished interview with Gregoria Sosa conducted by Adaljiza Sosa Riddell. Gracias a Adaljiza Sosa Riddell for providing me with a copy of this interview with her mother.

81. *Reader's Digest,* pp. 81–82. [Quote is from p. 82.]

82. Eusebia Buriel interview.

83. Cabeza de Baca, *We Fed Them,* p. 60.

84. Patricia Preciado Martin, *Songs My Mother Sang to Me* (Tucson: University of Arizona Press, 1992). The discussion that follows is taken from the foreword I wrote for the volume.

85. *Ibid.,* pp. 150–51.

86. *Ibid.,* p. 32.

87. *Ibid.,* pp. 199, 146, 161.

88. Thomas Sheridan, *Los Tucsonenses: The Mexican Community in Tucson, 1854–1941* (Tucson: University of Arizona Press, 1986), p. 147. Sheridan also provides an illuminating discussion of the interplay of religion and agrarian traditions. [See pages 159–63.]

89. Martin, *Songs My Mother Sang,* pp. 206, 111–12.

90. *Ibid.,* p. 129; Raquel Rubio Goldsmith, "Shipwrecked in the Desert: A Short History of the Mexican Sisters of the House of the Providence in Douglas, Arizona, 1927–1949," in *Women on the U.S.–Mexico Border: Responses to Change* (Boston: Allen and Unwin, 1987), pp. 177–95.

91. Martin, *Songs My Mother,* p. 207.

92. Richard White, *It's Your Misfortune and None of My Own: A New History of the American West* (Norman: University of Oklahoma Press, 1991), p. 288.

93. Ruiz interview; Deutsch, *No Separate Refuge,* pp. 155–58.

94. U.S. Congress, House, Committee on Immigration and Naturalization, *Western Hemisphere Immigration,* H.R. 8523, H.R. 8530, H.R. 8702, 71st. Cong., 2nd sess. (1930), p. 436.

95. *Ibid.*

96. Abraham Hoffman, *Unwanted Mexican Americans in the Great Depression: Repatriation Pressures, 1929–1939* (Tucson: University of Arizona Press, 1974), p. 29; Roy L. Garis, "The Mexican Invasion," *The Saturday Evening Post* (April 19, 1930): 43–44; Kenneth L. Roberts, "Wet and Other Mexicans," *The Saturday Evening Post* (February 4, 1928): 10–11, 137–38, 141–42, 146; Raymond G. Carroll, "The Alien on Relief," *The Saturday Evening Post* (January 11, 1936): 16–17, 100–103; Kenneth L. Roberts, "The Docile Mexican," *The Saturday Evening Post* (March 10, 1928): 39, 41, 165–66. [Quotes are from pp. 41 and 165, respectively.] For the best survey of Depression-era stereotypes, see Neil Betten and Raymond Mohl, "From Discrimination to Repatriation: Mexican Life in Gary, Indiana, During the Great Depression," in *The Chicano,* ed., Norris Hundley (Santa Barbara: ABC Clio Books, 1975), pp. 124–42.

97. George Horace Lorimer, "The Mexican Conquest," *The Saturday Evening Post* (June 22, 1929): 26.

98. Acuña, *Occupied America,* 2nd ed., pp. 138, 140–41; Camarillo, *Chicanos in California* (San Francisco: Boyd & Fraser, 1984), pp. 48–49; Hoffman, *Unwanted Mexican Americans,* pp. 43–46; Francisco Balderrama, *In Defense of La Raza: The Los Angeles Mexican Consulate and the Mexican Community, 1929–1936* (Tucson: University of Arizona Press, 1982), pp. 16–20; Betten and Mohl, "From Discrimination to Repatriation," pp. 132, 138–39.

99. George Sánchez, "The Rise of the Second Generation: The Mexican American Movement" (unpublished paper courtesy of the author), p. 10.

100. Balderrama and Rodríguez, *Decade of Betrayal,* pp. 121–22; 216. [Quote is from p. 216.]

101. Douglas Guy Monroy, "Mexicanos in Los Angeles, 1930–1941: An Ethnic Group in Relation to Class Forces" (Ph.D. dissertation, University of California, Los Angeles, 1978), p. 231. La migra refers to immigration agents.

102. Carey McWilliams, *North from Mexico: The Spanish-Speaking People of the United States* (Philadelphia: J.B. Lippincott, 1949; rpt. Greenwood Press, 1968), p. 193.

103. The two most comprehensive studies of deportation and repatriation to date include Francisco Balderrama and Raymond Rodríguez, *Decade of Betrayal* and Camille Guerin Gonzales, *Mexican Workers and American Dreams*.

104. Torres interview. She has lived in that house for over five decades.

105. Sánchez-Walker, "Woven Within My Grandmother's Braid," pp. 44, 46.

106. *Ibid,* p. 46.

107. *Ibid.,* pp. 60–62.

108. *Ibid.,* pp. 71–74. [Quotes are from pp. 73, 74, respectively.]

109. *Ibid.,* pp. 81, 83–87. The "nice ladies next door" who gave the daughters food in exchange for errands were probably prostitutes. With the exception of the youngest, all the children were native-born U.S. citizens and could cross the border with impunity. However, Petra and her infant could not; they would have to enter the United States with the services of a *coyote* or smuggler.

110. Betten and Mohl, "From Discrimination to Repatriation," p. 140.

111. Sánchez-Walker, "Woven Within My Grandmother's Braid," pp. 90–96. Back in the United States, Petra would bear two more children, one of whom would die as a toddler [pp. 100, 105–106].

112. Mario García, *Mexican Americans: Leadership, Ideology, and Identity, 1930–1960* (New Haven: Yale University Press, 1989), pp. 53–59; 62–83; García interview. For the more information on the Mexican American middle class, see Richard García, *Rise of the Mexican American Middle Class, San Antonio, 1929–1941* (College Station: Texas A&M University Press, 1991).

113. "Señora ____ ," Manuel Gamio field notes.

Chapter 2

1. Interview with Elsa Chávez, April 19, 1983, conducted by the author. *Note:* Elsa Chávez is a pseudonym used at the person's request.

2. Recent scholarship on Americanization programs aimed at Mexican communities includes George J. Sánchez, "'Go After the Women': Americanization and the Mexican Immigrant Woman, 1915–1929," in *Unequal Sisters: A Multicultural Reader in U.S. Women's History,* 2nd ed., eds. Vicki L. Ruiz and Ellen Carol DuBois (New York: Routledge, 1994), pp. 284–97; Sarah Deutsch, *No Separate Refuge: Culture, Class, and Gender on the Anglo-Hispanic Frontier in the American Southwest, 1880–1940* (New York: Oxford University Press, 1987), pp. 63–86; Gilbert González, *Chicano Education in the Era of Segregation* (Philadelphia: Balch Institute Press, 1990), pp. 30–61; Ruth Hutchinson Crocker, "Gary Mexicans and 'Christian Americanization': A Study in Cultural Conflict," in *Forging a Community: The Latino Experience in Northwest Indiana, 1919–1975,* eds. James B. Lane and Edward J. Escobar (Chicago: Cattails Press, 1987), pp. 115–34; Susan Yohn, *A Contest of Faiths: Missionary Women and Pluralism in the American Southwest* (Ithaca: Cornell University Press, 1995); Vicki L. Ruiz,

"Dead Ends or Gold Mines?: Using Missionary Records in Mexican American Women's History," *Frontiers: A Journal of Women's Studies*, 12:1 (1991): 33–56.

3. Pearl Idella Ellis, *Americanization Through Homemaking* (Los Angeles: Wetzel Publishing Co., 1929), preface [no page number].

4. *Ibid.*, p. 13.

5. M. Dorothy Woodruff, "Methodist Women Along the Mexican Border" (Women's Division of Christian Service pamphlet, ca. 1946) [part of an uncatalogued collection of documents housed at Houchen Community Center, El Paso, Texas; heretofore referred to as HF for Houchen Files]. This pamphlet provides brief descriptions of each of the twenty-one "centers of work" operated by Methodist missionaries. For a celebratory overview of Methodist women's missionary endeavors throughout the United States, see Noreen Dunn Tatum, *A Crown of Service* (Nashville, Tenn.: Parthenon Press, 1960).

6. Steven Seidman, ed., *Jürgen Habermas on Society and Politics: A Reader* (Boston: Beacon Press, 1989), p. 171.

7. Oscar J. Martínez, *The Chicanos of El Paso: An Assessment of Progress* (El Paso: Texas Western Press, 1980), pp. 6, 17.

8. Martínez, *Chicanos*, pp. 10, 29–33. Mario García meticulously documents the economic and social stratification of Mexicans in El Paso. See Mario T. García, *Desert Immigrants: The Mexicans of El Paso, 1880–1920* (New Haven: Yale University Press, 1981). *Note:* In 1960, the proportion of Mexican workers with high white-collar jobs jumped to 3.4 percent. [Martínez, *Chicanos*, p. 10.]

9. "South El Paso's Oasis of Care," *paso del norte*, Vol. I (September 1982): 42–43; Thelma Hammond, "Friendship Square," (Houchen Report, 1969) [HF]; "Growing with the Century" (Houchen Report, 1947) [HF].

10. García, *Desert Immigrants*. p. 145; Effie Stoltz, "Freeman Clinic: A Resume of Four Years Work" (Houchen Pamphlet, 1924) [HF]. It should be noted that Houchen Settlement sprang from the work of Methodist missionary Mary Tripp who arrived in South El Paso in 1893. However, it was not until 1912 that an actual settlement was established. ["South El Paso's Oasis of Care," p. 42].

11. Stoltz, "Freeman Clinic"; Hammond, "Friendship Square"; M. Dorothy Woodruff and Dorothy Little, "Friendship Square (Houchen Pamphlet, March 1949) [HF]; "Friendship Square" (Houchen Report, circa 1940s) [HF]; *Health Center* (Houchen Newsletter, 1943) [HF]; "Christian Health Service" (Houchen Report, 1941) [HF]; *El Paso Times*, October 20, 1945.

12. "Settlement Worker's Report" (Houchen Report, 1927) [HF]; Letter from Dorothy Little to E. Mae Young dated May 10, 1945 [HF]. Letter from Bessie Brinson to Treva Ely dated September 14, 1958 [HF]; Hammond, "Friendship Square"; Elmer T. Clark and Dorothy McConnell, "The Methodist Church and Latin Americans in the United States" (Board of Missions pamphlet, circa 1930s) [HF]. My very rough estimate is based on the documents and records to which I had access. I was not permitted to examine any materials then housed at Newark Hospital. The most complete statistics on utilization of services are for the year 1944 in the letter from Dorothy Little to E. Mae Young. *Note:* Because of the deportation and repatriation drives of the 1930s in which one-third of the Mexican population in the United States were either deported or repatriated, the Mexican population in El Paso dropped from 68,476 in 1930 to 55,000 in 1940. By 1960 it had risen to 63,796. [Martínez, *Chicanos*, p. 6]

13. *El Paso Herald Post*, March 7, 1961; *El Paso Herald Post*, March 12, 1961; "Community Centers" (Women's Division of Christian Service Pamphlet,

May 1963); *Funding Proposal* for Youth Outreach and Referral Report Project (April 30, 1974) [Private Files of Kenton J. Clymer, Ph.D.]; *El Paso Herald Post*, January 3, 1983; *El Paso Times*, August 8, 1983.

14. Letter from Tom Houghteling, Director, Houchen Community Center to the author December 24, 1990; Tom Houghteling, telephone conversation with the author, January 9, 1991.

15. Dorothy Little, "Rose Gregory Houchen Settlement" (Houchen Report, February 1942) [HF].

16. *Ibid.*; "Our Work at Houchen" (Houchen Report, circa 1940s) [HF]; Woodruff and Little, "Friendship Square"; Jennie C. Gilbert, "Settlements Under the Women's Home Missionary Society pamphlet, circa 1920s) [HF]; Clark and McConnell, "Latin Americans in the United States."

17. Anita Hernandez, "The Kindergarten" (Houchen Report, circa 1940s) [HF]; *A Right Glad New Year* (Houchen Newsletter, circa 1940s) [HF]; Little, "Rose Gregory Houchen Settlement"; "Our Work at Houchen"; Woodruff and Little, "Friendship Square." For more information on the Franciscans, see Ramón Gutiérrez, *When Jesus Came, the Corn Mothers Went Away: Marriage, Sexuality, and Power in New Mexico, 1500–1846* (Stanford: Stanford University Press, 1991).

18. Settlement Worker's Report (1927); Letter from Little to Young; letter from Brinson to Ely; *Friendship Square Calendar* (1949) [HF]; interview with Lucy Lucero, October 8, 1983, conducted by the author; Chávez interview; discussion following presentation, "Settlement Houses in El Paso," given by the author at the El Paso Conference on History and the Social Sciences, August 24, 1983, El Paso, Texas [tape of presentation and discussion is on file at the Institute of Oral History, University of Texas, El Paso].

19. Chávez interview; discussion following "Settlement Houses in El Paso." *Note:* The Catholic Church never established a competing settlement house. However, during the 1920s in Gary, Indiana, the Catholic diocese opened up the Gary-Alerding Settlement with the primary goal of Americanizing Mexican immigrants. The bishop took such action to counteract suspected inroads made by two local Protestant settlement houses. See Crocker, "Gary Mexicans," pp. 123–27.

20. "Christian Health Service"; "The Freeman Clinic and the Newark Conference Maternity Hospital" (Houchen Report, 1940) [HF]; *El Paso Times*, August 2, 1961; *El Paso Herald Post*, May 12, 1961. For more information on Americanization programs in California, see George J. Sánchez, "'Go After the Women,'" pp. 250–63. *Note:* The documents reveal a striking absence of adult Mexican male clients. The Mexican men who do appear are either Methodist ministers or lay volunteers.

21. Sánchez, "'Go After,'" pp. 250–83; Deutsch, *No Separate Refuge;* "Americanization Notes," *The Arizona Teacher and Home Journal*, 11:5 (January 1923): 26. *Note:* The Methodist and Presbyterian settlements in Gary, Indiana, also couched their programs in terms of "Christian Americanization." [Crocker, "Gary Mexicans," pp. 118–20]

22. "Settlement Worker's Report" (1927); *Friendship Square Calendar* (1949) [HF]; letter from Brinson to Ely; Chávez interview.

23. "News Clipping from *The El Paso Times*" (circa 1950s) [HF].

24. Clara Gertrude Smith, "The Development of the Mexican People in the Community of Watts" (M.A. thesis, University of Southern California, 1933), p. 104.

25. Rayna Green, "The Pocahontas Perplex," in *Unequal Sisters*, pp. 15–21.

26. Sánchez, "'Go After the Women,'" p. 260; *Newark-Houchen News*, September 1975. I agree with George Sánchez that Americanization programs created an overly rosy picture of American life. In his words: "Rather than providing Mexican immigrant women with an attainable picture of assimilation, Americanization programs could only offer these immigrants idealized versions of American life." [Sánchez, *loc. cit.*]

27. Little, "Rose Gregory Houchen Settlement."

28. García, *Desert Immigrants*, pp. 110–26; Paul S. Taylor, *Mexican Labor in the United States*, Vol. I (Berkeley: University of California Press, 1930, rpt. Arno Press, 1970), pp. 79, 205–206. [Quote is from Taylor, *Mexican Labor*, p. 79.]

29. Margarita B. Melville, "Selective Acculturation of Female Mexican Migrants," in *Twice a Minority: Mexican American Women*, ed. Margarita B. Melville (St Louis: C.V. Mosby, 1980), pp. 159–60; John García, "Ethnicity and Chicanos: Measurement of Etnic Identification, Identity, and Consciousness," *Hispanic Journal of Behavioral Sciences*, Vol. 4 (1982): 310–11. For an insightful, brief overview of Mexican-American ethnic identification, see David Gutiérrez, *Walls and Mirrors: Mexican Americans, Mexican Immigrants, and the Politics of Ethnicity in the Southwest, 1910–1986* (Berkeley: University of California Press, 1995), pp. 1–11.

30. Edward Soja, *Postmodern Geographies: The Reassertion of Space in Critical Social Theory* (New York and London: Verso Press, 1989), p. 23. Gracias a Matthew García for bringing this text to my attention.

31. "Settlement Worker's Report" (1927); Hernandez, "The Kindergarten" [HF]; *A Right Glad New Year*; Little, "Rose Gregory Houchen Settlement"; "Our Work at Houchen"; Woodruff and Little, "Friendship Square"; "South El Paso's Oasis of Care," *loc. cit.*; *El Paso Herald Post*, March 7, 1961; *El Paso Herald Post*, March 12, 1961; *El Paso Herald Post*, May 12, 1961.

32. C. S. Babbitt, "The Remedy for the Decadence of the Latin Race" (El Paso: El Paso Printing Company) (Presented to the Pioneers Association of El Paso, Texas, July 11, 1909, by Mrs. Babbitt, widow of the author), p. 55. Pamphlet courtesy of Jack Redman.

33. George J. Sánchez, *Becoming Mexican American: Ethnicity, Culture, and Identity in Chicano Los Angeles, 1900–1945* (New York: Oxford University Press, 1993), p. 156; Robert McLean, *That Mexican! As He Is, North and South of the Rio Grande* (New York: Fleming H. Revell Co., 1928), pp. 162–63, quoted in E. C. Orozco, *Republican Protestantism in Aztlán* (Santa Barbara: The Petereins Press, 1980), p. 162. Note: Immigration has frequently been linked with food from the "melting pot" of assimilation to the "salad bowl" of cultural pluralism. McLean's metaphor of Uncle Sam as a diner at the immigration cafe follows:

Fifty and one hundred years ago Uncle Sam accomplished some remarkable digestive feats. Gastronomically he was a marvel. He was not particularly choosy! Dark meat from the borders of the Mediterranean or light meat from the Baltic equally suited him, for promptly he was able to assimilate both, turning them into bone of his bone, and flesh of his flesh—But this chili con carne! Always it seems to give Uncle Samuel the heartburn; and the older he gets, the less he seems to be able to assimilate it. Indeed, it is a question whether chili is not a condiment, to be taken in small quantities rather than a regular article of diet. And upon this conviction ought to stand all the law . . . as far as the Mexican immigrant is concerned.

34. *Account Book for Rose Gregory Houchen Settlement* (1903–1913) [HF]; Hammond, "Friendship Square"; "Growing with the Century"; *El Paso Times,* September 5, 1975; Stoltz, "Freeman Clinic": Woodruff and Little, "Friendship Square"; *El Paso Times,* October 3, 1947; "Four Institutions. One Goal. The Christian Community" (Houchen pamphlet, circa early 1950s) [HF]; Houghteling conversation; "A City Block of Service" (Script of Houchen Slide Presentation, 1976) [HF]; *El Paso Times,* January 19, 1977; "Speech given by Kenton J. Clymer, Ph.D." (June 1975) [Clymer Files]; *El Paso Times,* May 23, 1975; *Newark-Houchen News,* September 1975. It should be noted that in 1904 local Methodist congregations did contribute much of the money needed to purchase the property on which the settlement was built. Local civic groups occasionally donated money or equipment and threw Christmas parties for Houchen children. [*Account Book; El Paso Herald Post,* December 14, 1951; *El Paso Times,* December 16, 1951]

35. Vernon McCombs, "Victories in the Latin American Mission" (Board of Home Missions pamphlet, 1935) [HF]; "Brillante Historia De La Iglesia 'El Buen Pastor' El Paso," *Young Adult Fellowship Newsletter,* December 1946 [HF]; Soledad Burciaga, "Yesterday in 1923" (Houchen Report, 1939) [HF].

36. This study is based on a limited number of oral interviews (five), but they represent a range of interaction with the settlement from playing on the playground to serving as the minister for El Buen Pastor. It is also informed by a public discussion of my work on Houchen held during an El Paso teachers' conference in 1983. Most of the educators who attended the talk had participated, to some extent, in Houchen activities and were eager to share their recollections. [C.f. note 13]. I am also indebted to students in my Mexican-American history classes when I taught at the University of Texas, El Paso, especially the reentry women, for their insight and knowledge.

37. Woodruff and Little, "Friendship Square"; Hammond, "Friendship Square"; *Greetings for 1946* (Houchen Christmas Newsletter, 1946) [HF]; Little, "Rose Gregory Houchen Settlement'; Soledad Burciaga, "Today in 1939" (Houchen Report, 1939) [HF]; "Our Work at Houchen"; "Christian Social Service" [Houchen Report, circa 1940s] [HF]; Interview with Fernando García, September 21, 1983, conducted by the author; *El Paso Times,* June 14, 1951; Lucero interview; Vicki L. Ruiz, "Oral History and La Mujer: The Rosa Guerrero Story," in *Women on the U.S.–Mexico Border: Responses to Change* (Boston: Allen and Unwin, 1987), pp. 226–27; *Newark-Houchen News,* September 1975.

38. Woodruff, "Mexican Women."

39. *Spanish-American Methodist News Bulletin,* April 1946 [HF]; Hammond, "Friendship Square"; McCombs, "Victories"; "El Metodismo en La Ciudad de El Paso," *Christian Herald,* July 1945 [HF]; "Brillante Historia"; "The Door: An Informal Pamphlet on the Work of the Methodist Church Among the Spanish-speaking of El Paso, Texas" (Methodist pamphlet, 1940) [HF]; "A City Block of Service" (script of slide presentation); García interview; Houghteling interview. *Note:* From 1932 to 1939, services for El Buen Pastor were held in a church located two blocks away from the settlement.

40. A. Ruth Kern, "There Is No Segregation Here," *Methodist Youth Fund Bulletin* (January–March 1953): 12 [HF].

41. *Ibid.;* "The Torres Family" (Houchen Report, circa 1940s) [HF]; interview with Estella Ibarra, November 11, 1982, conducted by Jesusita Ponce; Hazel Bulifant, "One Woman's Story" (Houchen Report, 1950) [HF]; "Our Work at Houchen."

42. Clara Sarmiento, "Lupe" (Houchen Report, circa 1950s) [HF].

43. Ibarra interview.

44. Bulifant, "One Woman's Story"; letter from Little to Young.

45. Deutsch, *No Separate Refuge,* pp. 64–66, 85–86; Sánchez, "'Go After the Women,'" pp. 259–61; Crocker, "'Gary Mexicans,'" p. 121.

46. Sarmiento, "Lupe." In her study, Ruth Crocker also notes the propensity of Protestant missionaries to focus their energies on children and the selective uses of services by Mexican clients. As she explained, "Inevitably, many immigrants came to the settlement, took what they wanted of its services, and remained untouched by its message." [Crocker, "Gary Mexicans," p. 122.]

47. *Newark-Houchen News,* September 1975.

48. Deutsch, *No Separate Refuge,* pp. 78–79; Ibarra interview; interview with Rose Escheverría Mulligan, Volume 27 of *Rosie the Riveter Revisited: Women and the World War II Work Experience,* ed. Sherna Berger Gluck (Long Beach: CSULB Foundation, 1983), p. 24.

49. Paul S. Taylor, "Women in Industry," field notes for *Mexican Labor in the United States,* Bancroft Library, University of California, Berkeley, Box 1. *Note:* Referring to Los Angeles, two historians have argued that "Mexicans experienced segregation in housing in nearly every section of the city and its outlying areas." [Antonio Ríos-Bustamante and Pedro Castillo, *An Illustrated History of Mexican Los Angeles* (Los Angeles: UCLA Chicano Studies Research Center, 1986), p. 135.]

50. Interview with Alicia Mendeola Shelit, Volume 37, *Rosie the Riveter Revisited,* p. 32; Mulligan interview, p. 14. Anthropologist Ruth Tuck noted that Euro-Americans also employed the term "Spanish" to distinguish individuals "of superior background or achievement." [Ruth Tuck, *Not with the Fist* (New York: Harcourt, Brace and Co., 1946; rpt. Arno Press, 1974), pp. 142–43.]

51. Interview with Alma Araiza García, March 27, 1993, conducted by the author.

52. Tuck, *Not with the Fist,* p. 133.

53. *Friendship Square Calendar* (1949); Beatrice Fernandez, "Day Nursery" (Houchen Report, circa late 1950s) [HF].

54. "Friendship Square" (Houchen pamphlet, circa 1950s) [HF]; letter to Houchen Girl Scouts from Troop 4, Latin American Community Center, Alpine, Texas, May 18, 1951 [HF].

55. *A Right Glad New Year;* News clipping from the *El Paso Times* (circa 1950s); "Our Work at Houchen"; Little, "Rose Gregory Houchen Settlement"; "Anglo Settlement Worker's Journal" (entry for December 1952) [HF].

56. *Newark-Houchen News,* September 1975; Sarmiento, "Lupe."

57. Peggy Pascoe, *Relations of Rescue: The Search for Moral Authority in the American West, 1874–1939* (New York: Oxford University Press, 1990), pp. 112–39; Woodruff, "Methodist Women."

58. *Datebook for 1926* (entry: Friday, September 9, 1929) (Settlement Worker's private journal) [HF]; "Brillante Historia"; "Report and Directory of Association of Church Social Workers, 1940" [HF]; "May I Come in? (Houchen brochure, circa 1950s) [HF]; "Friendship Square" (Houchen pamphlet, 1958) [HF]; Mary Lou López, "Kindergarten Report" (Houchen Report, circa 1950s) [HF]; Sarmiento, "Lupe"; "Freeman Clinic and Newark Hospital" (Houchen pamphlet, 1954) [HF]; *El Paso Times,* June 14, 1951; "Houchen Day Nursery" (Houchen pamphlet, circa 1950s) [HF]; *El Paso Times,* September 12, 1952.

59. Chávez interview; Martha González, interview with the author, October 8, 1983; Lucero interview; *Newark-Houchen News,* September 1974.

60. Letter from Little to Young; "The Door"; Woodruff and Little, "Friendship Square."

61. "Houchen Day Nursery"; "Life in a Glass House" (Houchen Report, circa 1950s) [HF].

62. *Program* for First Annual Meeting, Houchen Settlement and Day Nursery, Freeman Clinic, and Newark Conference Maternity Hospital (January 8, 1960) [HF]. It should be noted that thirty years later, there seems to be a shift back to original settlement ideas. Today, Houchen Community has regularly scheduled bible studies. [Letter from Houghteling to the author.]

63. *Program* for Houchen production of "Cinderella" [HF]; letter from Brinson to Ely. For more information on LULAC, see Mario T. García, *Mexican Americans: Leadership, Ideology, and Identity, 1930–1960* (New Haven: Yale University Press, 1989).

64. Bulifant, "One Woman's Story"; *News from Friendship Square* (Spring newsletter, circa early 1960s) [HF].

65. My understanding and application of the ideas of Jürgen Habermas have been informed by the following works. Jürgen Habermas, *Moral Consciousness and Communicative Action,* trans. Christian Lenhardt and Sherry Weber Nicholsen, introd. Thomas McCarthy (Cambridge: MIT Press, 1990); Seidman, ed., *Jürgen Habermas on Society and Politics;* Nancy Fraser, *Unruly Practices: Power, Discourse, and Gender in Contemporary Social Theory* (Minneapolis: University of Minnesota Press, 1989); Seyla Benhabib and Drucilla Cornell, "Introduction: Beyond the Politics of Gender," in *Feminism as Critique,* eds. Seyla Benhabib and Drucilla Cornell (Minneapolis: University of Minnesota Press, 1987).

66. As an example of this typology, see Mario García, *Mexican Americans,* pp. 13–22, 295–302. Richard Griswold del Castillo touches on the dynamic nature of Mexican culture in *La Familia: Chicano Families in the Urban Southwest, 1848 to the Present* (Notre Dame: University of Notre Dame Press, 1984).

67. Escheverría Mulligan interview, p. 17.

68. Louise Año Nuevo Kerr, "The Chicano Experience in Chicago, 1920–1970" (Ph.D. dissertation, University of Illinois, Chicago Circle, 1976), p. 104.

Chapter 3

1. Interview with María Ybarra, December 1, 1990, conducted by David Pérez.

2. For colonial New Mexico, Ramón Gutiérrez convincingly demonstrates how family honor was tied, in part, to women's *vergüenza* (literally, shame or virginity). See Ramón Gutiérrez, "Honor, Ideology, and Class Gender Domination in New Mexico, 1690–1846," *Latin American Perspectives* 12 (Winter 1985): 81–104, and Ramón Gutiérrez, *When Jesus Came, the Corn Mothers Went Away: Power and Sexuality in New Mexico, 1500–1846* (Stanford: Stanford University Press, 1990). I contend that since mothers and elder female relatives played major roles in enforcing chaperonage, strict supervision of daughters related more to what I term "familial oligarchy" than to patriarchal control.

3. Donna R. Gabaccia, *From Sicily to Elizabeth Street* (Albany: SUNY Press,

1984); Sydney Stahl Weinberg, "The Treatment of Women in Immigration History: A Call for Change," in *Seeking Common Ground: Multidisciplinary Studies of Immigrant Women in the United States*, ed. Donna Gabaccia (Westport, Conn.: Greenwood Press, 1992), pp. 3–22; Stuart and Elizabeth Ewen, *Channels of Desire* (New York: McGraw-Hill, 1982); Elizabeth Ewen, *Immigrant Women in the Land of Dollars* (New York: Monthly Review Press, 1985); Andrew Heinze, *Adapting to Abundance* (New York: Columbia University Press, 1990); Susan A. Glenn, *Daughters of the Shtetl* (Ithaca: Cornell University Press, 1990).

4. Interview with Erminia Ruiz, February 18, 1993, conducted by the author. In 1919, *True Story Magazine* began an industry of romance and scandal magazines; by 1929, this publication had over two million subscribers. See Joanne J. Meyerowitz, *Women Adrift: Independent Wage Earners in Chicago, 1880–1930* (Chicago: University of Chicago Press), pp. 129–33.

5. Interview with Jesusita Torres, January 8, 1993, conducted by the author.

6. George Lipsitz, *Time Passages: Collective Memory and American Popular Culture* (Minneapolis: University of Minnesota Press, 1990), p. 13; Nirwan Dewanto, "American Kitsch and Indonesian Culture: A Sketch," paper presented at the Third Annual International Symposium: American Studies in the Asia-Pacific Region, "The Impact of American Popular Culture on Social Transformation in Asian Countries" (April 3, 1992) Tokyo, Japan. For the most comprehensive examination of the impact of mass culture on immigrants and their children, see Lizbeth Cohen, *Making a New Deal: Industrial Workers in Chicago, 1919–1939* (Cambridge: Cambridge University Press, 1990).

7. Ruth Tuck, *Not with the Fist* (New York: Harcourt, Brace and Co., 1946; rpt. Arno Press, 1974), pp. 185–88; Vicki L. Ruiz, "Oral History and La Mujer: The Rosa Guerrero Story," in *Women on the U.S.–Mexico Border: Responses to Change* (Boston: Allen and Unwin, 1987)," pp. 226–27; interview with Belen Martínez Mason, Volume 23 of *Rosie the Riveter Revisited: Women and the World War I Work Experience*, ed. Sherna Berger Gluck (Long Beach: CSULB Foundation, 1983), pp. 24–25; Ruiz interview (1993); Interview with Ruby Estrada, August 4, 1981, conducted by María Hernández, "The Lives of Arizona Women" Oral History Project (On File, Special Collections, Hayden Library, Arizona State University, Tempe, Arizona), p. 6. I gratefully acknowledge ASU archivist Christine Marín for providing me with this interview. For a scholarly overview of the educational experiences of Spanish-speaking children in southwestern schools, see Gilbert González, *Chicano Education in the Era of Segregation* (Philadelphia: Balch Institute Press, 1990).

8. Interview with Mary Luna, Vol. 20 of *Rosie the Riveter Revisited*, p. 10. During the 1940s, bilingual education appeared as an exciting experiment in curriculum reform. See "First Regional Conference on the Education of the Spanish-Speaking People in the Southwest," ed. George I. Sánchez (December 1945) reprinted in *Aspects of the Mexican American Experience*.

9. Luna interview, p. 9.

10. Valerie J. Matsumoto, *Farming the Home Place: A Japanese American Community in California, 1919–1982* (Ithaca: Cornell University Press, 1993); Valerie J. Matsumoto, "Growing Up with Jeanette MacDonald and Dear Deidre: Nisei Girls in the 1930s" (unpublished paper courtesy of the author); Judy Yung, *Unbound Feet: A Social History of Chinese Women in San Francisco* (Berkeley: University of California Press, 1995).

11. Mauricio Mazón's *The Zoot Suit Riots* (Austin: University of Texas Press, 1984) and the Luis Valdez play and feature film, *Zoot Suit* provide examples of the literature on *pachucos*. A doctoral student at Princeton University, Eduardo Pagán is completing a dissertation on pachucos and the politics of race during World War II.

12. I would like to introduce these women by grouping them geographically. María Fierro, Rose Escheverria Mulligan, Adele Hernández Milligan, Beatrice Morales Clifton, Mary Luna, Alicia Mendeola Shelit, Carmen Bernal Escobar, Belen Martínez Mason, and Julia Luna Mount grew up in Los Angeles. Lucy Acosta and Alma Araiza García came of age in El Paso and Erminia Ruiz in Denver. Representing the rural experience are María Arredondo, and Jesusita Torres (California), María Ybarra (Texas), and Ruby Estrada (Arizona). As a teenager, Eusebia Buriel moved with her family from Silvis, Illinois, to Riverside, California. *Note:* Of the seventeen full-blown life histories, nine are housed in university archives, seven as part of the *Rosie the Riveter* collection at California State University, Long Beach. I appreciate the generosity and longstanding support of Sherna Gluck who has given me permission to use excerpts from the *Rosie* interviews. This sample also does not include oral interviews found in published sources.

13. The age breakdowns for the fourteen interviewees are as follows: nine were born between 1908 and 1919 and eight between 1920 and 1926. This sample includes some who were chaperoned during the 1920s and others who were chaperoned during the thirties and forties. As a result, the sample does not a represent a precise generational grouping, but instead gives a sense of the pervasiveness and persistence of unremitting supervision.

14. Estrada interview, pp. 2, 15, 17, 19. *Note:* Most families were nuclear, rather than extended, although kin usually (but not always) resided nearby. Carmen Bernal Escobar and Alma Araiza García grew up in single-parent households with extended kin present, Carmen reared by her mother, Alma by her father.

15. George J. Sánchez, "'Go After the Women': Americanization and the Mexican Immigrant Woman 1915–1929," in *Unequal Sisters: A Multicultural Reader in U.S. Women's History*, 2nd ed., eds. Vicki L. Ruiz and Ellen Carol DuBois (New York: Routledge, 1994), p. 285.

16. F. Scott Fitzgerald, *Flappers and Philosophers* (London: W. Collins Sons and Co., Ltd., 1922), pp. 209–46; Emory S. Bogardus, *The Mexican in the United States* (Los Angeles: University of Southern California Press, 1934), p. 741; Martínez Mason interview, p. 44. During the 1920s, Mexican parents were not atypical in voicing their concerns over the attitudes and appearance of their "flapper adolescents." A general atmosphere of tension between youth and their elders existed—a generation gap that cut across class, race, ethnicity, and region. See Paula Fass, *The Damned and the Beautiful: American Youth in the 1920's* (New York: Oxford University Press, 1977).

17. Interview with Alicia Mendeola Shelit, Volume 37 of *Rosie the Riveter*, p. 18; Paul S. Taylor, *Mexican Labor in the United States, Volume II* (Berkeley: University of California Press, 1932), pp. 199–200; Interview with María Fierro, Volume 12 of *Rosie the Riveter*, p. 10. Changing clothes at school is not peculiar to our mothers and grandmothers. As a high school student in the early 1970s, I was not allowed to wear the fashionable micro-mini skirts. But I bought one anyway. I left home in a full dirndl skirt with a flowing peasant blouse, but once I arrived at

school, I would untie the skirt (which I would then dump in my locker) to reveal the mini-skirt I had worn underneath.

18. Manuel Gamio, *Mexican Immigration to the United States* (Chicago: University of Chicago Press, 1930; rpt. Arno Press, 1969), p. 89. The verse taken from "Las Pelonas" in the original Spanish follows:

> Los paños colorados
> Los tengo aborrecidos
> Ya hora las pelonas
> Los usan de vestidos.
> Las muchachas de S. Antonio
> Son flojas pa'l metate
> Quieren andar pelonas
> Con sombreros de petate.
> Se acabaron las pizcas,
> Se acabó el algodón
> Ya andan las pelonas
> De puro vacilón.

19. Taylor, *Mexican Labor, Vol. II*, pp. vi–vii.

20. Rodolfo F. Acuña, *Community Under Siege: A Chronicle of Chicanos East of the Los Angeles River, 1945–1975* (Los Angeles: UCLA Chicano Studies Publications, 1984), pp. 278, 407–408, 413–14, 418, 422; *FTA News*, May 1, 1945; interview with Carmen Bernal Escobar, June 15, 1986, conducted by the author. For an example of the promotion of a beauty pageant, see issues of *La Opinion*, June–July 1927.

21. Escobar interview, 1986.

22. Sherna B. Gluck, *Rosie the Riveter Revisited: Women, The War and Social Change* (Boston: Twayne Publishers, 1987), pp. 81, 85.

23. The best elaboration of this phenomenon can be found in Roland Marchand, *Advertising the American Dream: Making Way for Modernity, 1920–1940* (Berkeley: University of California Press, 1985).

24. For examples, see *La Opinion*, September 26, 1926; May 14, 1927; June 5, 1927; September 9, 1929; January 15, 1933; January 29, 1938. Lorena Chambers is currently writing a dissertation focusing on the gendered representations of the body in Chicano cultural narratives. I thank her for our wonderful discussions.

25. Vicki L. Ruiz, "'Star Struck': Acculturation, Adolescence, and Mexican American Women, 1920–1940" in *Small Worlds: Children and Adolescents in America*, eds. Elliot West and Paula Petrik (Lawrence: University of Kansas Press, 1992): 61–80; Roberto R. Treviño, *"Prensa Y Patria*: The Spanish-Language Press and the Biculturation of the Tejano Middle Class, 1920–1940," *The Western Historical Quarterly*, Vol. 22 (November 1991): 460.

26. *La Opinion*, September 29, 1929.

27. *Hispano-America*, July 2, 1932. Gracias a Gabriela Arredondo for sharing this advertisement with me, one she included in her seminar paper, "'Equality' for All: Americanization of Mexican Immigrant Women in Los Angeles and San Francisco Through Newspaper Advertising, 1927–1935 (M.A. seminar paper, San Francisco State University, 1991).

28. *La Opinion*, June 5, 1927; *La Opinion*, February 8, 1938.

29. Richard A. García, *Rise of the Mexican American Middle Class: San Antonio, 1929–1941* (College Station: Texas A&M Press, 1991), pp. 118–19; Treviño, *"Prensa Y Patria,"* pp. 459–60.

30. For examples, see *La Opinion,* September 23, 1926; *La Opinion,* September 24 1926; *La Opinion,* September 27, 1926; *La Opinion,* September 30, 1926; *La Opinion,* June 4, 1927; *La Opinion,* February 27, 1931; and *La Opinion,* August 17, 1931.

31. For an elaboration of this theme, see Ruiz, "'Star Struck'." The quote is taken from *La Opinion,* March 2, 1927.

32. Ewen and Ewen, *Channels of Desire,* pp. 95–96.

33. Lipsitz, *Time Passages,* p. 16.

34. The struggles young Mexican-American women faced just to talk freely with men and attend the movies unchaperoned stand in stark contrast to their Euro-American peers who had passed first base and were headed toward greater liberties, like having a drink in a bar without tainting their reputations. See Mary Murphy, "Bootlegging Mothers and Drinking Daughters: Gender and Prohibition in Butte, Montana," *American Quarterly,* 46:2 (June 1994): 174–94.

35. Martínez Mason interview, pp. 29–30; Ybarra interview; Escobar interview; Fierro interview, p. 15; Estrada interview, pp. 11–12; interview with Erminia Ruiz, July 30, 1990, conducted by the author; Ruiz interview, 1993, conducted by the author; interview with Alma Araiza García, March 27, 1993, conducted by the author. Chaperonage was also common in Italian immigrant communities. Indeed, many of the same conflicts between parents and daughters had surfaced a generation earlier among Italian families on the East Coast, although in some communities chaperonage persisted into the 1920s. See Kathy Peiss, *Cheap Amusements: Working Women and Leisure in Turn-of-the Century New York* (Philadelphia: Temple University Press, 1986), pp. 69–70, 152.

36. Interview with Adele Hernández Milligan, Volume 26 of *Rosie the Riveter,* p. 17.

37. Evangeline Hymer, "A Study of the Social Attitudes of Adult Mexican Immigrants in Los Angeles and Vicinity: 1923" (M.A. thesis, University of Southern California, 1924; rpt. San Francisco: R and E Research Associates, 1971), pp. 24–25. Other ethnographies that deal with intergenerational tension include Helen Douglas, "The Conflict of Cultures in First Generation Mexicans in Santa Ana, California" (M.A. thesis, University of Southern California, 1928) and Clara Gertrude Smith, "The Development of the Mexican People in the Community of Watts" (M.A. thesis, University of Southern California, 1933).

38. Escobar interview, 1986; Estrada interview, pp. 11, 13; interview no. 653 with Lucy Acosta conducted by Mario T. García, October 28, 1982 (on file at the Institute of Oral History, University of Texas, El Paso), p. 17. I wish to thank Rebecca Craver, coordinator of the Institute of Oral History, for permission to use excerpts from the Acosta interview.

39. Estrada interview, p. 12; Shelit interview, p. 9; Antonio Ríos-Bustamante and Pedro Castillo, *An Illustrated History of Mexican Los Angeles, 1781–1985* (Los Angeles: Chicano Studies Research Center, UCLA, 1986), p. 153.

40. Paul S. Taylor, "Women in Industry," field notes for his book, *Mexican Labor in the Unites States, 1927–1930,* Bancroft Library, University of California, 1 box; Richard G. Thurston, "Urbanization and Sociocultural Change in a Mexican-American Enclave" (Ph.D. dissertation, University of California, Los Ange-

les, 1957; rpt. R and E Research Associates, 1974), p. 118; Bogardus, *The Mexican,* pp. 28–29, 57–58. *Note:* Paul S. Taylor's two-volume study, *Mexican Labor in the United States,* is considered the classic ethnography on Mexican Americans during the interwar period. A synthesis of his field notes, "Women in Industry," has been published. See Taylor, "Mexican Women in Los Angeles Industry in 1928," *Aztlán,* 11 (Spring 1980): 99–131.

41. Martínez Mason interview p. 30; Ruiz interviews (1990, 1993); Thomas Sheridan, *Los Tucsonenses* (Tucson: University of Arizona Press, 1986), pp. 131–32.

42. Sheridan, *Los Tucsonenses, loc. cit.*

43. Interview with Beatrice Morales Clifton, Volume 8 of *Rosie the Riveter,* pp. 14–15.

44. Smith, "The Development of the Mexican People," p. 47.

45. Ybarra interview. Ethnographies by Smith, Thurston, and Douglas refer to elopement as a manifestation of generational tension.

46. Discussion following my presentation of "The Flapper and the Chaperone" at the Riverside Municipal Museum, May 28, 1995. Comment provided by Rose Medina, co-curator with Vincent Moses of the museum's special exhibition: "Nuestros Antepasados: Riverside's Mexican American Community, 1917–1950."

47. Shelit interview, pp. 9, 24, 30; Ruiz interviews (1990, 1993); Escobar interview; García interview; Martínez Mason interview p. 30; Hernández Milligan interview, pp. 27–28; interview with María Arredondo, March 19, 1986, conducted by Carolyn Arredondo; Taylor notes.

48. Interview with Julia Luna Mount, November 17, 1983, by the author; Fierro interview, p. 18; Luna interview, p. 29; Ruiz interview (1993); Gregorita Rodríguez, *Singing for My Echo* (Santa Fe: Cota Editions, 1987), p. 52; Martínez Mason interview, p. 62.

49. "Elisa Morales," interview by Luis Recinos, April 16, 1927, Biographies and Case Histories II folder, Manuel Gamio Field Notes, Bancroft Library, University of California; Taylor, *Mexican Labor,* Vol. II, pp. vi–vii; Gamio, *Mexican Immigration,* p. 89. The corrido "El Enganchado" in Volume two of *Mexican Labor* offers an intriguing glimpse into attitudes toward women and Americanization.

50. Ruth Alexander in her study of wayward girls in New York City makes an important point about this balancing of boundaries among teenagers.

> Using a variety of strategies, the great majority of adolescent girls and young women must have negotiated America's urban terrain in relative safety, enjoying and inventing a sexualized lifestyle while acknowledging the limits of their freedom and acting to protect themselves from social stigma or state action.

[Ruth Alexander, "'The Only Thing I Wanted Was Freedom': Wayward Girls in New York, 1900–1930," in *Small Worlds,* p. 294.]

51. Ruiz interview (1993).

52. Carmen and Diego are pseudonyms used to ensure the privacy of the family. Carmen's oral interview is in the author's possession.

53. Discussion following my presentation, of "The Flapper and the Chaperone," May 28, 1995. Comment provided by B.V. Meyer.

54. Douglas Monroy, "An Essay on Understanding the Work Experiences of Mexicans in Southern California, 1900–1939, *Aztlán,* 12 (Spring 1981): 70. *Note:* Feminist historians have also documented this push for autonomy among the

daughters of European immigrants. In particular, see Peiss, *Cheap Amusements,* Glenn, *Daughters of the Shtetl,* E. Ewen, *Immigrant Women;* and Alexander, "The Only Thing I Wanted Was Freedom." See also Meyerowitz, *Women Adrift.*

55. Heller Committee for Research in Social Economics of the University of California and Constantine Panuzio, *How Mexicans Earn and Live,* University of California Publications in Economics, XIII, No. 1, Cost of Living Studies V (Berkeley: University of California, 1933), pp. 11, 14, 17; Taylor notes; Luna Mount interview; Ruiz interviews (1990, 1993); Shelit interview, p. 9. For further delineation of the family wage economy, see Louise A. Tilly and Joan W. Scott, *Women, Work, and Family* (New York: Holt, Rinehart, and Winston, 1978).

56. These observations are drawn from my reading of the seventeen oral interviews and the literature on European immigrant women.

57. Luna Mount interview.

58. García interview.

59. Rev. F. X. Lasance, *The Catholic Girl's Guide and Sunday Missal* (New York: Benziger Brothers, 1905), Esther Pérez Papers, Cassiano-Pérez Collection, Daughters of the Republic of Texas Library at the Alamo, San Antonio, Texas, pp. 279–80. I have a 1946 reprint edition passed down to me by my older sister who had received it from our mother.

60. Gutiérrez, "Honor, Ideology," pp. 88–93, 95–98.

61. Estrada interview, p. 12. Focusing on the daughters of European immigrants, Elizabeth Ewen has written that "the appropriation of an urban adolescent culture" served as "a wedge against patriarchal forms of social control." This holds true, to some degree, for the women profiled here. But, for Mexican Americans, the underlying ideological assumption was familial oligarchy rather than patriarchy. See Ewen and Ewen, *Channels of Desire,* p. 95.

62. Lasance, *Catholic Girl's Guide,* pp. 249–75. [Quote is on p. 270.]

63. *Ibid.,* p. 271.

64. George J. Sánchez, *Becoming Mexican American: Ethnicity, Culture, and Identity in Chicano Los Angeles, 1900–1945* (New York: Oxford University Press, 1993), p. 167; Mary Helen Ponce, *Hoyt Street* (Albuquerque: University of New Mexico Press, 1993), pp. 258, 266–71. [Quote is taken from p. 258.] "Las vistas" is slang for the movies.

65. Ponce, *Hoyt Street,* p. 266; Margo McBane, "Tale of Two Cities: A Comparative Study of the Citrus Heartlands of Santa Paula and LaVerne" (unpublished paper courtesy of the author), pp. 15–16. Discussion following my presentation of "The Flapper and the Chaperone," May 28, 1995. Comment provided by Rose Medina. *Jamaicas* are church bazaars or festivals.

66. Interview with Eusebia Buriel, January 16, 1995, conducted by the author; interview with Ray Buriel, December 21, 1994, conducted by the author.

67. Patricia Preciado Martin, *Songs My Mother Sang to Me* (Tucson: University of Arizona Press, 1992), pp. 19–20.

68. Interview with Rose Escheverria Mulligan, Volume 27 of *Rosie the Riveter,* p. 24.

69. Taylor notes; Monroy, "An Essay on Understanding," p. 70; Rosalinda González, "Chicanas and Mexican Immigrant Families 1920–1940: Women's Subordination and Economic Exploitation," in *Decades of Discontent: The Women's Movement, 1920–1940,* eds. Lois Scharf and Joan M. Jensen (Westport, Conn.: Greenwood Press, 1983), p. 72; Vicki L. Ruiz, *Cannery Women, Cannery Lives: Mexican Women, Unionization, and the California Food Processing Industry,*

1930–1950 (Albuquerque: University of New Mexico Press, 1987), pp. 10–12, 17–18; Ruiz interview (1990); John D'Emilio and Estelle B. Freedman, *Intimate Matters: A History of Sexuality in America* (New York: Harper & Row, 1988), pp. 233–35, 239–41. *Note: Intimate Matters* provides a thought-provoking analysis of sexual liberalism during the interwar period.

70. *La Opinion,* May 9, 1927.

71. Tuck, *Not with the Fist,* p. 134.

72. Only two (Ruby Estrada and Alma García) achieved a solid, consistent middle-class standard of living. Six of the thirteen California women took their place at the shop floor in the aerospace, electronics, apparel, and food-processing industries. Two became secretaries and one a sales clerk at K-Mart. Jesusita Torres has been a farm and nursery worker since the age of nine. The remaining three narrators can be considered *amas de casa* or homemakers.

73. Among the husbands of the California women, many were skilled workers in the aerospace industry and the highest occupation for a spouse was firefighter. *Note:* While intermarriage among Mexican American is generally conceptualized in terms of Euro-American husbands, in the Imperial Valley of California, substantial numbers of Mexican women married Punjabi men. See Karen Isaksen Leonard, *Making Ethnic Choices: California's Punjabi Mexican Americans* (Philadelphia: Temple University Press, 1992).

74. Margaret Clark, *Health in the Mexican American Culture* (Berkeley: University of California Press, 1959), p. 20.

75. Albert Camarillo, "Mexican American Urban History in Comparative Ethnic Perspective," Distinguished Speakers Series, University of California, Davis (January 26, 1987); Rodolfo Acuña, *Occupied America: A History of Chicanos,* 2nd ed. (New York: Harper & Row, 1981), pp. 310, 318, 323, 330–31; Shelit interview, p. 15.

76. Paul S. Taylor, *Mexican Labor in the United States,* Vol. I (Berkeley: University of California Press, 1930; rpt. Arno Press, 1970), pp. 221–24; Arredondo interview; Ruiz interviews (1990, 1993).

77. Ruiz interview (1993).

78. *Ibid.*

79. *Ibid.*

80. *Ibid.* At the age of twenty, Erminia would marry an Anglo Army Air Corps lieutenant after a brief courtship. Neither family attended the ceremony held in a Denver church rectory. Her husband would later describe Erminia as "the work horse of the family." After her marriage, she regularly sent her mother money and later she and two of her sisters would take turns caring for their elderly mother.

81. Ruiz, *Cannery Women,* pp. 18–19, 35–36. For a splendid study on the nexus between acculturation and wage work among contemporary Mexican immigrant women, see Pierrette Hondagneu-Sotelo, *Gendered Transitions: Mexican Experiences of Immigrations* (Berkeley: University of California Press, 1994).

82. Nicolás Kanellos, *A History of Hispanic Theatre in the United States* (Austin: University of Texas Press, 1990), pp. 39, 42–43. As María Teresa Montoya explained, "'nuestros compatriotas no son ni mexicanos ya, ni americanos.'" [p. 39].

83. Acosta interview; Tuck, *Not with the Fist,* pp. 126–27; Thurston, "Urbanization," pp. 109, 117–19; Ruiz interviews (1990, 1993).

84. Bernice Zamora, "Pueblo, 1950," in *Infinite Divisions: An Anthology of Chicana Literature,* eds. Tey Diana Rebolledo and Eliana Rivero (Tucson: University of Arizona Press, 1993), p. 315.

85. Ruiz interviews (1990, 1993); personal experience of author. I should note that during the early 1960s, I served as my sister's chaperone whenever she had a date at the drive-in movie.

Chapter 4

1. Letter from Carey McWilliams dated October 3, 1937, to Louis Adamic, *Adamic File,* Carton 1, Carey McWilliams Collection, Older material [call number 1243], Special Collections, University of California, Los Angeles, California.
2. *Ibid.*
3. *Ibid.;* Interview with Dorothy Ray Healey, January 21, 1979, conducted by the author.
4. Founded in 1927, the League of United Latin American Citizens (LU-LAC) is the oldest civil rights organization among Mexican Americans. Women's participation in LULAC will be addressed later in this chapter.
5. Born in Guatemala, Luisa Moreno brought Latinos together in labor and political causes from organizing Puerto Rican and Cuban seamstresses in New York City to orchestrating the Spanish-speaking Peoples Congress. The daughter of a prominent New Mexican ranching family, Concha Ortiz y Pino was the first Hispana elected to the state of House of Representatives. Although having different political orientations, both women were advocates for bilingual education. I will address their contributions later in the chapter.
6. Cynthia E. Orozco, "Beyond Machismo, La Familia, and Ladies Auxiliaries: A Historiography of Mexican-Origin Women's Participation in Voluntary Associations and Politics in the United States, 1870–1990," in *Renato Rosaldo Lecture Series Monograph,* Vol. 10 (Tucson: University of Arizona Mexican American Studies and Research Center, 1992–93), p. 39.
7. Ibid., p. 40. Though not covering union organization, Orozco's overview essay provides the most prescient historical survey of Mexican women's community activism.
8. *Los Angeles Times,* April 25, 1903. Despite their efforts, the strike collapsed. Although not mentioning women's involvement, historian Charles Wollenberg has published the most extensive treatment of Mexican rail workers' first attempt to unionize. See Charles Wollenberg, "Working on El Traque: The Pacific Electric Strike of 1903," in *The Chicano,* ed., Norris Hundley (Santa Barbara: ABC Clio, 1978), pp. 96–107.
9. E. P. Thompson, *The Making of the English Working Class* (New York: Vintage Books, 1963), p. 807.
10. Sam Kushner, *Long Road to Delano* (New York: International Publishers, 1975), pp. 64–65, 72; Healey interview; "Elizabeth Nicholas: Working in the California Canneries" [interview conducted by Ann Baxandall Krooth and Jaclyn Greenberg], *Harvest Quarterly,* Nos. 3–4 (September–December, 1976): 21; Cletus E. Daniel, *Bitter Harvest: A History of California Farmworkers, 1870–1941* (Ithaca: Cornell University Press, 1981), pp. 110–11.
11. Carey McWilliams, *Factories in the Field: The Story of Migratory Farm Labor in California* (Boston: Little, Brown, 1935; rpt. Santa Barbara: Peregrine, 1971), pp. 214–29; Daniel, *Bitter Harvest,* pp. 219–20; Walter J. Stein, *California and the Dust Bowl Migration* (Westport, Conn.: Greenwood Press, 1973), p. 224; Ramón Chacón, "The 1933 San Joaquin Valley Cotton Strike: Strikebreaking Activities in California Agriculture," in *Work, Family, Sex Roles, Language,* eds.

Mario Barerra, Alberto Camarillo, and Francisco Hernández (Berkeley: To-natiuth-Quinto Sol, 1980), p. 34.

12. Devra Weber, *Dark Sweat, White Gold: California Farm Workers, Cotton, and the New Deal* (Berkeley: University of California Press, 1994), p. 91. Weber provides the most nuanced and detailed examination of the 1933 San Joaquin Valley Cotton Strike. See Chapter three, "As the faulting of the earth . . . The Strike of 1933," pp. 79–111.

13. Chacón, "1933 Strike," pp. 36–38, 43–62; Pledge Sheet (circa October 1933), *CAWIU File,* Simon J. Lubin Collection, Bancroft Library, University of California, Berkeley; Weber, *Dark Sweat,* pp. 100–102; 107–109; Rudolfo Acuña, *Occupied America,* 2nd ed. (New York: Harper & Row, 1981), pp. 222–26. [Quote is excerpted from p. 223.]

14. Weber, *Dark Sweat,* pp. 94–97. [Quote is from p. 96.]

15. Ronald W. López, "The El Monte Berry Strike of 1933," *Aztlán,* Vol. 1 (Spring 1970): 103–104; Charles S. Spaulding, "The Mexican Strike at El Monte, California," *Sociology and Social Research,* Vol. 18 (September–August, 1933–1934): 572–73; Interview with Jesusita Torres, January 8, 1993, conducted by the author.

16. Spaulding, "The Mexican Strike," p. 571; López, "The El Monte Strike," p. 103; Acuña, *Occupied America,* p. 220; *Los Angeles Times,* September 27, 1992; Torres interview.

17. Acuña, *Occupied America,* pp. 220–21; López, "The El Monte Berry Strike," pp. 105–106, 108–109.

18. *Los Angeles Times,* September 27, 1992; López, "The El Monte Berry Strike," p. 106; Torres interview. Zenaida "Sadie" Castro made quite an impression on Holguin who described Castro as follows: "She was one of the few people who spoke English back then. . . . She had some Indian blood in her. She was feisty, and she would boss her husband around. There was no bumbling around her." [*Los Angeles Times,* September 27, 1992.]

19. Torres interview; Spaulding, "The Mexican Strike of 1933," p. 578; López, "The El Monte Berry Strike," pp. 108–109. López contends that the public ploy was not very successful because the strikers verbally and physically intimidated the bargain hunters.

20. López, "The El Monte Berry Strike," pp. 108–109; Acuña, *Occupied America,* p. 221; Spaulding, "The Mexican Strike," pp. 578–80; Torres interview.

21. John Steinbeck, "Their Blood Is Strong." Simon J. Lubin Society pamphlet (1938), pp. 19–21, 31–33.

22. Healey interview.

23. *San Francisco News,* July 22, 1940. For slightly different takes on the degrees of worker consciousness among Mexican field hands during the 1930s, see Weber, *Dark Sweat, White Gold;* Camille Guerin-Gonzales, *Mexican Workers and American Dreams: Immigration, Repatriation, and California Farm Labor, 1900–1939* (New Brunswick, N.J.: Rutgers University Press, 1994); Victor B. Nelson-Cisneros, "UCAPAWA and Chicanos in California: The Farm Worker Period, 1937–1940," *Aztlán,* Vol. 7 (Fall 1976): 453–77.

24. U.S. Congress, Senate, Committee on Education and Labor, *Hearings Before a Subcommittee of the Senate Committee on Education and Labor Violations of Free Speech and Rights of Labor Part 70,* pp. 25736. [commonly known as the LaFollette Committee Hearings]. The Associated Farmers were bankrolled by powerful corporate interests in California, including the Pacific Gas and Electric

Company. For more information on this vigilante group, see Vicki L. Ruiz, *Cannery Women, Cannery Lives* (Albuquerque: University of New Mexico Press, 1937), pp. 50–55; Weber, *Dark Sweat, White Gold,* pp. 118–23, 189–98.

25. U.S. Senate Committee on Education and Labor, *Hearings Before a Subcommittee of the Senate Committee on Education and Labor Violations of Free Speech and Rights of Labor Part 71,* pp. 26394–97.

26. "Blood on the Cotton," UCAPAWA pamphlet (1939), p. 1; U.S. Senate Committee on Education and Labor, *Hearings Before a Subcommittee of the Senate Committee on Education and Labor Violations of Free Speech and Rights of Labor Part 51,* pp. 18653, 18838–39, 18913.

27. U.S. Senate Committee on Education and Labor, *Hearings on the Rights of Labor Part 71,* pp. 26262, 26270–71, 26275–77, 26282–303; Clarke Alexander Chambers, "A Comparative Study of Farm Organizations in California During the Depression Years, 1929–1941" (Ph.D. dissertation, University of California, Berkeley, 1950), pp. 158, 160; Weber, *Dark Sweat, White Gold,* pp. 196–98. [Quote is from p. 198.]

28. *Agricultural Bulletin,* November 15, 1939. This was the monthly newsletter of the National Council to Aid Agricultural Workers.

29. Ruiz, *Cannery Women, Cannery Lives,* pp. 55–57; Weber, *Dark Sweat, White Gold,* pp. 195, 198–99, 207–10.

30. Acuña, *Occupied America,* p. 234; Robert Garland Landolt, "The Mexican-American Workers of San Antonio, Texas" (Ph.D. dissertation, University of Texas, 1965), pp. 228–35; Julia Kirk Blackwelder, *Women of the Depression: Caste and Culture in San Antonio* (College Station: Texas A&M Press, 1984), p. 143. [Quote is from Acuña, p. 234.]

31. "San Antonio: the Cradle and the Coffin of Texas Liberty" (Texas Civil Liberties Union pamphlet, 1938), George Lambert Collection, Labor Archives, University of Texas at Arlington, Arlington, Texas, p. 3.

32. *Ibid.,* pp. 5–8; *The Daily Worker,* June 28, 1938; Irene Ledesma, "Texas Newspapers and Chicana Workers' Activism, 1919–1974," *Western Historical Quarterly,* Vol. 26 (Autumn 1995): 317–20; Roberto R. Calderón and Emilio Zamora, "Manuela Solis Sager and Emma Tenayuca: A Tribute," in *Chicana Voices: Intersections of Class, Race, and Gender,* eds. Teresa Córdova et al. (Austin: University of Texas, Center for Mexican American Studies, 1986), pp. 33, 37.

33. Zaragosa Vargas, "'La Pasionaria de Tejas': Emma Tenayuca, 1930s Mexican American Communist Labor Organizer of San Antonio, Texas" (unpublished paper, courtesy of the author), p. 43. *The Daily Worker* described Emma Tenayuca: "The beautiful young girl looked hardly big enough or old enough to have become a leader of the Mexican workers' fight for economic and civil rights in Texas." [*The Daily Worker,* June 28, 1938.]

34. Acuña, *Occupied America,* pp. 234–36; Interview with Luisa Moreno, August 12–13, 1977, conducted by Albert Camarillo; interview with Luisa Moreno, July 27, 1978, conducted by the author; *Decision of the Board of Arbitration in San Antonio Pecan Strike,* San Antonio, Texas (April 13, 1938), Lambert Collection, UTA Labor Archives. *Note:* Henderson's decision to send Luisa Moreno "infuriated" Emma Tenayuca, who reluctantly stepped aside. As a result, the working relationship between the two women proved tenuous and tense. [Vargas, "La Pasionaria," pp. 31–32; Moreno interviews, 1977 and 1978.] Julia Kirk Blackwelder argues that "Tenayuca and the Workers Alliance wrestled with the

CIO for control of the workers"; but Tenayuca "agreed to take a back seat" to ensure the strike's success. Blackwelder does not mention Luisa Moreno in her narrative. [Blackwelder, *Women of the Depression*, p. 148.]

35. Landolt, "The Mexican American Workers," pp. 228, 234–35.

36. Ledesma, "Texas Newspapers and Chicana Activism," p. 321; letter from Maizie Tamez to Donald Kobler, June 3, 1942, Lambert Collection, UTA Labor Archives.

37. Interview with Carmen Bernal Escobar, February 11, 1979, conducted by the author. For more information, see Chapter 1 of my monograph, *Cannery Women, Cannery Lives.* [Vicki L. Ruiz, *Cannery Women, Cannery Lives: Mexican Women, Unionization, and the California Food Processing Industry, 1930-1950* (Albuquerque: University of New Mexico Press, 1987).]

38. See Chapters 2 and 4 of *Cannery Women.* Quotes are from Escobar interview and Moreno interview, 1978, respectively.

39. See Chapters 2 and 4 of *Cannery Women, Cannery Lives.*

40. See Chapter 4 of *Cannery Women, Cannery Lives* or the article version of the Cal San Strike—Vicki L. Ruiz, "A Promise Fulfilled: Mexicana Cannery Workers of Southern California," in *The Pacific Historian*, 30 (Summer 1986): 50–61.

41. *Ibid.*

42. Sara Evans has defined "social space" as an area "within which members of an oppressed group can develop an independent sense of worth in contrast to their received definitions as second-class or inferior citizens." [Sara M. Evans, *Personal Politics: The Rise of Women's Liberation in the Civil Rights Movement and the New Left* (New York: Vintage Books, 1980), p. 219.]

43. Escobar interview.

44. Ruiz, *Cannery Women, Cannery Lives,* pp. 78–80; Carmen R. Chávez, "Coming of Age During the War: Reminiscences of an Albuquerque Hispana," *New Mexico Historical Review*, 70:4 (October 1995): 396–97; interview with Alicia Mendeola Shelit, Volume 37 of *Rosie the Riveter Revisited: Women and the World War I Work Experience*, ed. Sherna Berger Gluck (Long Beach: CSULB Foundation, 1983), pp. 13, 52–55. See also Sherna Berger Gluck, *Rosie the Riveter Revisited: Women, the War, and Social Change* (Boston: Twayne Publishers, 1987).

45. Interview with Julia Luna Mount, November 17, 1983, conducted by the author.

46. *Ibid.*

47. Patricia Rae Adler, "The 1943 Zoot Suit Riot: Brief Episode in a Long Conflict," in *An Awakening Minority: The Mexican-Americans*, ed. Manuel P. Servín quoted in Mauricio Mazón, *The Zoot Suit Riots: The Psychology of Symbolic Annihilation* (Austin: University of Texas Press, 1984), p. 64; Carmen Lomas Garza, "Pachuca with a Razor Blade" (original in the collection of Sonía Saldívar Hull). *Note:* Of course, pachucas represented only one segment of the Mexican-American adolescent population.

48. Carey McWilliams, *North from Mexico: The Spanish-Speaking People of the United States* (Philadelphia: J.B. Lippincott, 1949; rpt. Greenwood Press, 1968), p. 257; *La Opinion*, August 26, 1942, as cited in Jesus Francisco Malaret, "Of Pachucos, Zoot Suiters, and the Press" (unpublished seminar paper, University of California, Davis, 1992), pp. 90–91.

49. McWilliams, *North from Mexico*, pp. 257–58.

50. Stark's statement appears in the *Los Angeles Evening Herald Express,* May 20, 1943. For more information on the Zoot Suit Riots, see McWilliams, *North from Mexico* and Mazón, *The Zoot Suit Riots.*

51. My interpretation is not meant to excuse the tiny minority of women who engaged in delinquent behavior, but to bring out the voices of those unfairly castigated on the basis of ethnicity, gender, youth, and personal appearance.

52. William H. Chafe, *The American Woman: Her Changing Social, Economic, and Political Roles, 1920–1970* (New York: Oxford University Press, 1972), pp. 137–43, 146; Mario Barrera, *Race and Class in the Southwest* (Notre Dame: University of Notre Dame Press, 1979), pp. 131, 140–45.

53. Interview with Josefina Fierro de Bright, August 7, 1977, conducted by Albert Camarillo; William M. Tuttle, Jr., *"Daddy's Gone to War": The Second World War in the Lives of American Children* (New York: Oxford University Press, 1993), pp. 166–67. In retrospect, Negrete felt remorse at taunting her Japanese-American neighbors.

54. Gilbert González, *Chicano Education in an Era of Segregation* (Philadelphia: Balch Institute Press, 1990), pp. 136–56; "Episode 3: Blow-Outs" of the PBS documentary *CHICANO!* (Galán Productions, 1996); Ruiz, *Cannery Women, Cannery Lives,* pp. 112–23; "Moreno, Luisa," by Vicki L. Ruiz, *Encyclopedia of the American West,* eds., Charles Phillips and Alan Axelrod (New York: Macmillan, 1996), p. 1030. In 1944, UCAPAWA changed its name to the more manageable Food, Tobacco, Agricultural, and Allied Workers of America (FTA). For more information on the demise of the union, see Chapter 6 of *Cannery Women, loc. cit.*

55. Michael Wilson and Deborah Silverton, *Salt of the Earth* (Old Westbury: Feminist Press, 1978), pp. 93–154; *Salt of the Earth,* feature film, Paul Jarrico, producer (1954); *New York Times,* May 3, 1982. [Quotes are from p. 122.] Gracias a Tom Miller for providing me with this news clipping.

56. Wilson and Rosenfeldt, *Salt of the Earth,* pp. 124–25, 137–38, 140–41.

57. *Ibid.,* pp. 93–95, 130–32, 171–76, 183–84; *New York Times,* May 3, 1982. See Herbert Biberman, *Salt of the Earth: The Story of a Film* (Boston: Beacon Press, 1965). [Quotes are from pp. 171 and 176.] Although the film was roundly criticized and played in only a few American theaters in 1954, it received recognition in Europe and enjoyed a rebirth in the United States during the late 1960s and early 1970s, the heyday of both the Chicano Movement and Women's Liberation.

58. *New York Times,* May 3, 1982. In her commentary, Rosenfelt offers a more complicated reading of gender relations among Mexican mining families. She contends that, for many women, their families "reverted back to the old way" in the years following the strike. [Wilson and Rosenfelt, *Salt of the Earth,* pp. 142–45.]

59. Albert Camarillo, *Chicanos in a Changing Society: From Mexican Pueblos to American Barrios in Santa Barbara and Southern California, 1848–1930* (Cambridge: Harvard University Press, 1979), pp. 147–48; Juan Gómez-Quiñones, *Roots of Chicano Politics, 1600–1940* (Albuquerque: University of New Mexico Press, 1994), pp. 312–14; Thomas Sheridan, *Los Tucsonenses: The Mexican Community in Tucson, 1854–1941* (Tucson: University of Arizona Press, 1986), pp. 107–108; Mario T. García, *Memories of Chicano History: The Life and Narrative of Bert Corona* (Berkeley: University of California Press, 1994), p. 294. See also Roberto R. Calderón, "Union, Paz y Trabajo: Laredo's Mutual Aid Soci-

eties, 1890s" in *Historical Scholarship in Texas: Proceedings of the Mexicans in Texas History Conference,* eds. Emilio Zamora, Cynthia E. Orozco, and Rudolfo Rocha (Austin: University of Texas Press, forthcoming). *Note: Mutualistas* originated among workers in Mexico during the 1850s. In addition, they exemplified the types of mutual aid societies created by immigrants across the globe. [Calderón, "Union, Paz y Trabajo," p. 3; Sheridan, *Los Tucsonenses,* p. 108.]

60. Sheridan, *Los Tucsonenses,* pp. 108, 112; Mario T. García, *Desert Immigrants: The Mexicans of El Paso, 1880–1920* (New Haven: Yale University Press, 1980), p. 226.

61. Camarillo, *Chicanos in a Changing Society,* pp. 148–51; Richard A. García, *Rise of the Mexican Middle Class, San Antonio, 1929–1941* (College Station: Texas A & M Press, 1991), pp. 110–12.

62. Orozco, "Beyond Machismo," pp. 48–51; Camarillo, *Chicanos in a Changing Society,* pp. 150–51.

63. Gilbert González, *Labor and Community: Mexican Citrus Worker Villages in a Southern California County, 1900–1950* (Urbana: University of Illinois Press, 1994), p. 82; Zaragosa Vargas, *Proletarians of the North: A History of Mexican Industrial Workers in Detroit and the Midwest, 1917–1933* (Berkeley: University of California Press, 1993), p. 155; Gómez-Quiñones, *Roots of Chicano Politics,* p. 312; Calderón, "Union, Paz y Trabajo," pp. 11–12 (typescript courtesy of author).

64. David G. Gutiérrez, *Walls and Mirrors: Mexican Americans, Mexican Immigrants, and the Politics of Ethnicity* (Berkeley: University of California Press, 1995), pp. 96–99. [Quote is on p. 97.]

65. F. Arturo Rosales, "Shifting Self-Perception and Ethnic Consciousness Among Mexicans in Houston, 1908–1946," *Aztlán,* Vol. 16 (1985): 76.

66. Sheridan, *Los Tucsonenses,* p. 167; Gómez-Quiñones, *Roots of Chicano Politics,* p. 312.

67. Francisco E. Balderrama and Raymond Rodriguez, *Decade of Betrayal: Mexican Repatriation in the 1930s* (Albuquerque: University of New Mexico Press, 1995), p. 43; Rosales, "Shifting Self-Perceptions," p. 74; Sarah Deutsch, *No Separate Refuge: Culture, Class, and Gender on an Anglo-Hispanic Frontier in the American Southwest, 1850–1940* (New York: Oxford University Press, 1987), p. 154; Vargas, *Proletarians of the North,* pp. 150, 152; Acuña, *Occupied America,* p. 310.

68. Francisco E. Balderrama, *In Defense of La Raza: The Los Angeles Mexican Consulate and the Mexican Community, 1929 to 1936* (Albuquerque: University of New Mexico Press, 1982), p. 39; García, *Rise of the Mexican American,* pp. 11–112.

69. Louise Año Nuevo Kerr, "The Chicano Experience in Chicago, 1920–1970" (Ph.D. dissertation, University of Illinois, Chicago Circle, 1976), p. 49.

70. *The Lemon Grove Incident,* documentary produced by Paul Espinosa (1985); Balderrama, *In Defense of La Raza,* pp. 58–61; Robert R. Alvarez, Jr., *Familia: Migration and Adaptation in Baja and Alta California, 1800–1975* (Berkeley: University of California Press, 1987), pp. 152–55. [Quote is from p. 154.] *Note:* Balderrama cites a 1931 survey that indicated "that more than 80 percent of the school districts in southern California enrolled Mexicans and Mexican Americans in segregated schools" [p. 56].

71. *The Lemon Grove Incident;* Balderrama, *In Defense of La Raza,* pp. 60–61; Alvarez. *Familia,* pp. 152–55. [Quote is from p. 154.]

72. Orozco, "Beyond Machismo," p. 49; Sheridan, *Los Tucsonenses*, p. 168.

73. The pattern of voluntarist politics among Mexican women appears congruent with activities by African-American and Anglo women during the same period. See Sara M. Evans, *Born For Liberty: A History of Women in America* (New York: The Free Press, 1989).

74. Gutiérrez, *Walls and Mirrors*, p. 6.

75. *Ibid.*

76. Benjamin Márquez, *LULAC: The Evolution of a Mexican American Political Organization* (Austin: University of Texas Press, 1993), pp. 17–38; Gutiérrez, *Walls and Mirrors*, pp. 74–87. [Quote is from p. 77.]

77. Mario T. García, *Mexican Americans: Leadership, Ideology, and Identity, 1930–1960* (New Haven: Yale University Press, 1989), p. 35; Gutiérrez, *Walls and Mirrors*, p. 84.

78. *Ibid*, p. 78.

79. *Ibid,*; García, *Mexican Americans*, pp. 56–57; Márquez, *LULAC*, pp. 53–55.

80. Gutiérrez, *Walls and Mirrors*, pp. 84–85; García, *Rise of the Mexican American Middle Class*, p. 63; Márquez, *LULAC*, pp. 30–33. [Quote is from p. 33.]

81. Gutiérrez, *Walls and Mirrors*, pp. 80–81.

82. Orozco, "Beyond Machismo," pp. 53–55; Cynthia E. Orozco, "The Origins of the League of United Latin American Citizens (LULAC) and the Mexican American Civil Rights Movement in Texas with an Analysis of Women's Political Participation in a Gendered Context, 1910–1929" (Ph.D. dissertation, University of California, Los Angeles, 1992). [Quote is from Orozco, "Beyond Machismo," p. 55.]

83. "Alicia Dickerson Montemayor (1902–1989)" in *Las Mujeres: Mexican American/Chicana Women* (National Women's History Project curriculum booklet, 1991), p. 15; Orozco, "Beyond Machismo," p. 53. [Quote is from "Alicia Dickerson Montemayor," *loc. cit.*] For further reference, see Cynthia E. Orozco, "Alicia Dickerson Montemayor: The Feminist Challenge to the League of United Latin American Citizens, Family Ideology, and Mexican American Politics in Texas in the 1930s," in *Writing the Range: Race, Class, and Culture in the Women's West*, eds. Susan Armitage and Elizabeth Jameson (Norman: University of Oklahoma Press, 1997, pp. 435–56).

84. Elizabeth Salas, "Ethnicity, Gender and Divorce: Issues in the 1922 Campaign by Adelina Otero-Warren for the U.S. House of Representatives," *New Mexico Historical Review*, 70:4 (October 1995): 367, 369–71; Charlotte Whaley, *Nina Otero-Warren of Santa Fe* (Albuquerque: University of New Mexico Press, 1994), pp. 8, 11–15, 27, 47, 74–75. While in New York, Otero Warren also kept house for her brother, a student at Columbia.

85. Whaley, *Nina Otero-Warren*, pp. 77–100. [Quotes are on pp. 86 and 94, respectively.]

86. *Ibid.*, p. 94. Note: Joan Jensen touches on Otero Warren's lobbying skills and on her later impact on voter turnout among Hispanic women in "'Disenfranchisement Is a Disgrace': Women and Politics in New Mexico, 1900–1940," in *New Mexico Women: Intercultural Perspectives*, eds. Joan Jensen and Darlis Miller (Albuquerque: University of New Mexico Press, 1986), pp. 301–31.

87. Whaley, *Nina Otero-Warren*, pp. 95–98; Salas, "Ethnicity, Gender, and Divorce," pp. 373, 376–77, 379 [Quote is from p. 377.]

88. Salas, "Ethnicity, Gender, and Divorce," pp. 375–80. Note: During the

1922 election, Democrat Soledad Chávez de Chacón was elected Secretary of State.

89. *Ibid.*, p. 372; Whaley, *Nina Otero-Warren,* pp. 116–21, 130, 148–49, 152; Tey Diana Rebolledo and Eliana S. Rivero, *Infinite Divisions: An Anthology of Chicana Literature* (Tucson: University of Arizona Press, 1993), pp. 17–19; Tey Diana Rebolledo, *Women Singing in the Snow: A Cultural Analysis of Chicana Literature* (Tucson: University of Arizona Press, 1995), pp. 34–37. For an interesting discussion on the politics of Spanish identity during this period, see Philip B. Gonzales, "Spanish Heritage and Ethnic Protest in New Mexico: The Anti-Fraternity Bill of 1933," *New Mexico Historical Review,* Vol. 61 (October, 1986): 281–98. I would also note that Maricela Chávez is conducting insightful research on Hispana cultural resistance through writing. Maricela R. Chávez, "Nuevomexicana Cultural Invention as Cultural Resistance: Fabiola Cabeza de Baca, Cleofas M. Jaramillo, and Nina Otero Warren, 1880–1955" (graduate seminar paper, Arizona State University, 1996).

90. "Resume of Concha Ortiz y Pino de Kleven," Collection #457 Concha Ortiz y Pino de Kleven Collection, Box 1, Folder 1, University of New Mexico, Special Collections, University of New Mexico General Libraries. [Hereafter the collection will be referred to as OPK-UNM] "Concha Ortiz y Pino de Kleven (b. 1912)," in *Las Mujeres,* p. 19; newsclippings, n.d., *Private Scrapbook of Concha Ortiz y Pino de Kleven,* Box 2, OPK-UNM; interview with Concha Ortiz y Pino de Kleven, September 16, 1988, conducted by the author; newsclippings on Concha Ortiz y Pino, n.d., Collection #303 Women in New Mexico Collection, Box 1, Folder 9, UNM; *Albuquerque Journal,* March 22, 1941. Concha Ortiz y Pino was also the first woman legislator in the United States to hold the position of majority whip.

91. Connie Meyer, "Concha Ortiz y Pino de Kleven," *Century: A Southwest Journal of Observation and Opinion* (July 15, 1981), OPK-UNM, Box 1, Folder 15. Concha Ortiz y Pino de Kleven's Spanish colonial heritage is mentioned in every featured news story. For example, the text under a photo caption reads as follows: EL GRECO MIGHT HAVE painted this highbred face. La Señora Doña Concha Ortiz y Pino de Kleven with her aquiline features, green eyes, and sable hair is the descendant of a Spanish family that came as colonizers 400 years ago to what is now the State of New Mexico." [*The Los Angeles Times* (n.d.), OPK-UNM, Box 1, Folder 12.]

92. Ortiz y Pino de Kleven interview.

93. *Ibid.*

94. "New Mexico's Aristocratic Lady" (newsclipping from *Albuquerque Journal,* n.d.), OPK-UNM, Box 1, Folder 13.

95. "Vignette: Concha Ortiz y Pino de Kleven," *The Santa Fean* (December 1974–1975): 18–19; Concha Ortiz y Pino de Kleven, "Biographical Statement" (3 pp., typescript), OPK-UNM, Box 1, Folder 1; "Her achievements reflect proud heritage" (newsclipping, n.d.), OPK-UNM, Box 1, Folder 13; "Long Cultural Heritage Backs Her Interest in Fine Arts" (newsclipping, n.d.), OPK-UNM, Box 1, Folder 14; newsclipping (n.d.), *Private Scrapbook;* Ortiz y Pino de Kleven interview. On the first page of her biographical statement, she explained how the school functioned. "And the money earned . . . went to the artisan until he could afford to purchase his own materials. Then he was able to buy and sell as he pleased, but continued to operate in the school aiding the less skilled."

96. Concha Ortiz y Pino's Voting Record (field notes compiled by Thomas Jaehn based on legislative documents housed at the New Mexico State Records

Center and Archive, Santa Fe); *Santa Fe New Mexican,* February 5, 1937; *Albuquerque Journal,* March 22, 1941. Again, I would like to thank Thomas Jaehn for graciously sharing his research materials.

97. Ortiz y Pino de Kleven interview. During our interview, she stated, "I was not a feminist. I was a humanist. I saw no difference between men and women." In 1977, she was a bit more blunt with a local reporter. "I refuse any mail that comes addressed 'Ms'. I want my identity and my 'Mrs'." ["Concha Ortiz y Pino de Kleven: Civic Leader" (newsclipping, July 14, 1977), OPK-UNM, Box 1, Folder 13.]

98. "Resume of Concha Ortiz y Pino de Kleven"; Joan M. Jensen, "Pioneers in Politics," *El Palacio,* Vol. 92 (Summer–Fall, 1986): 16; Ortiz y Pino de Kleven interview. Concha remains very proud of her efforts with regard to civil service reform.

> But the one thing I did fight for, I wanted to see the merit system. I had seen so many of the girls who had gone to work in the capital and had ended up with babies. . . . It was my dream and it did come true."

[Ortiz y Pino de Kleven interview.]

99. Newsclippings (ca. January–March, 1941), *Private Scrapbook;* "Resume of Concha Ortiz y Pino"; *Outlook,* July 14, 1982, OPK-UNM, Box 1, Folder 13.

100. Newspaper Engagement Announcement (ca. 1943), OPK-UNM, Box 1, Folder 13; *Albuquerque Tribune,* June 6, 1983; *Albuquerque Journal,* December 4, 1966; "Her achievements mark proud heritage." *Note:* Ortiz y Pino de Kleven was also involved in establishing a high school equivalency program in the state prison.

101. "New Mexico's Aristocratic Lady"; *Outlook,* July 14, 1982; "Resume of Concha Ortiz y Pino de Kleven"; Ortiz y Pino de Kleven interview, *New York Times,* April 9, 1982. [Quote is from the *New York Times.*]

102. Moreno interviews (1977, 1978).

103. Moreno interviews (1977, 1978); Albert Camarillo, *Chicanos in California* (San Francisco: Boyd and Fraser, 1984), p. 61; Gutiérrez, *Walls and Mirrors,* p. 111; Acuña, *Occupied America,* p. 317.

104. Gutiérrez, *Walls and Mirrors,* pp. 112–15; Sánchez, *Becoming Mexican American,* pp. 246–49; Camarillo, *Chicanos in California,* pp. 63–64.

105. Gómez-Quiñones, *Roots of Chicano Politics,* pp. 383–87; Gutiérrez, *Walls and Mirrors,* pp. 112–15; Camarillo, *Chicanos in California,* pp. 63–64; García, *Mexican Americans,* pp. 154–57.

106. Sánchez, *Becoming Mexican American,* p. 246.

107. "Moreno, Luisa *loc. cit.*"; Moreno interviews (1977, 1978); interview with Luisa Moreno, September 4–6, 1979, conducted by the author; interview with Luisa Moreno, August 3, 1984, conducted by the author. Her parents christened her Blanca Rosa Rodríguez López, but when she began organizing in the United States she took the name Luisa Moreno. In 1927, her critically acclaimed collection of poems entitled *El Venedor de Cocuyos* was published in Mexico City (copy in author's possession).

108. García, *Mexican Americans,* p. 155; Fierro de Bright interview. [Quote is from interview.] For more information on the popularity of Magón and his party among Mexican women, see Emma M. Pérez, "A La Mujer: A Critique of the Partido Liberal Mexicano's Gender Ideology on Women," in *Between Borders,* pp. 459–82.

109. Fierro de Bright interview; García, *Mexican Americans,* pp. 155–56.

John Bright would later be blacklisted as a member of the Hollywood Ten.

110. Fierro de Bright interview; Moreno interviews (1977, 1978, 1979); García, *Mexican Americans,* pp. 158–59. [Quote is from p. 159].

111. Moreno interviews (1977, 1978, 1979); García, *Mexican Americans,* p. 158. [Quote is from p. 158.]

112. Moreno interviews (1977, 1984); García, *Mexican Americans,* pp. 164–65; Fierro de Bright interview. [Quote is from Fierro de Bright interview.]

113. Luisa Moreno, "Caravans of Sorrow" presentation given at the panel on Deportation and the Right of Asylum at the Fourth Annual Conference of the American Committee for the Protection of the Foreign Born (Washington, D.C., March 3, 1940), McWilliams Collection, Box 1, Folder 1.

114. Louisa [sic] Moreno, "Non Citizen Americans of the South West" (New York: ACPFB pamphlet, 1942), La Gente exhibit, Colorado Historical Society Library, Denver, Colorado; García, *Mexican Americans,* pp. 159–60.

115. The American G.I. Forum, a post-World War II organization, has worked for the rights of Mexican-American veterans and their families and, like LULAC, has kept a political distance from immigrants. Asociación Nacional México-Americana (ANMA) grew out of the labor movement and addressed local issues such as police brutality. The Community Service Organization (CSO), established in California in 1947, blended voter registration drives, citizenship programs, and civil rights activism. Like El Congreso, CSO sought to address the needs of both Mexican immigrants and Mexican Americans. For more information, see Gutiérrez, *Walls and Mirrors* and García, *Mexican Americans.* For women's activism in CSO, see Margaret Rose, "The Community Service Organization, 1947–1962," in *Not June Cleaver: Women and Gender in Postwar America, 1945–1960* (Philadelphia: Temple University Press, 1994), pp. 177–200. In addition, Chicana historian Linda Apodaca is currently writing a book focusing on women in the CSO.

116. Victor Becerra, "The Untold Story of Chávez Ravine" (unpublished paper), history research holdings, Chicano Studies Library, University of California, Los Angeles, p. 12. Becerra further observed: "The case of Chávez Ravine is more than a story of baseball in Los Angeles, or the struggle of Ravine residents to save their homes and their community. Rather, it is a statement about City Planning and how capitalists create programs such as urban renewal and redevelopment to legitimize [their] exappropriation of resources for personal profit" (p. 13).

117. Acuña, *Occupied America,* p. 350; Thomas P. Carter, *Mexican Americans in School: A History of Educational Neglect* (Princeton: College Entrance Examination Board, 1970), p. 23.

Chapter 5

1. Inés Hernández, "Sara Estela Ramírez: Sembradora," *Legacy,* 6:1 (1989): 22. The original in Spanish follows:

> Surge! Surge a la vida, a la actividad, a la
> belleza de vivir realmente; perso surge radiante
> y poderosa, bella de cualidades, esplendente
> de virtudes, fuerte de energias.

[*La Cronica,* April 9, 1910.]

2. For information on Sara Estela Ramírez, see Hernández, "Sara Estela Ramírez," pp. 13–26; Emilio Zamora, Jr., "Sara Estela Ramírez: Una Rosa Roja en el Movimiento," in *Mexican Women in the United States: Struggles Past and Present,* eds. Magdalena Mora and Adelaida R. Del Castillo (Los Angeles: UCLA Chicano Studies Research Center Publications, 1980), pp. 163–69.

3. Jose E. Limón, "El Primer Congreso Mexicanista de 1911: A Precursor to Contemporary Chicanismo," *Aztlán,* 5:1–2 (Spring and Fall, 1974): 91–99; Emilio Zamora, *The World of the Mexican Worker in Texas* (College Station: Texas A&M Press, 1993), pp. 97–99; Alfredo Mirandé and Evangelina Enríquez, *La Chicana: The Mexican-American Woman* (Chicago: University of Chicago Press, 1979), p. 223; Marta Cotera, *Diosa y Hembra: The History and Heritage of Chicanas in the U.S.* (Austin: Information Systems Development, 1976), pp. 68–71 [Quote is from p. 70.]

4. Examples include Carlos Muñoz, Jr., *Youth, Identity, and Power: The Chicano Movement* (New York: Verso, 1989), p. 160; Marguerite V. Marín, *Social Protest in an Urban Barrio: A Study of the Chicano Movement, 1966–1974* (Lanham, Md.: University Press of America, 1991) pp. 135–39; Blanche Linden-Ward and Carol Hurd Green, *American Women in the 1960s: Changing the Future* (New York: Twayne Publishers, 1993); Sara Evans, *Born for Liberty* (New York: Free Press, 1989). The poem "La Nueva Chicana" by Viola Correa can be found in the Appendix. It is quoted from Teresa Córdova, "Roots and Resistance: The Emergent Writings of Twenty Years of Chicana Feminist Struggle," in *Handbook of Hispanic Cultures in the United States: Sociology,* ed. Félix Padilla (Houston: Arte Publico Press, 1994), p. 182.

5. Adela de la Torre and Beatríz Pesquera, eds., *Building with Our Hands: New Directions in Chicana Studies* (Berkeley: University of California Press, 1993); Tey Diana Rebolledo and Eliana S. Rivero, eds., *Infinite Divisions: An Anthology of Chicana Literature* (Tucson: University of Arizona Press, 1993). Ruben Martínez has referred to Chicano culture as "swirls of cultural contradictions." [Ruben Martínez, "The Role of the Intellectual Chicano/Latino in our Communities," talk given at The Claremont Graduate Humanities Center, October 6, 1993.]

6. Elizabeth Martínez, "Chingón Politics Die Hard: Reflections on the First Chicano Activists Reunion," *Z Magazine* (April 1990): 48.

7. Mario T. García, *Mexican Americans: Leadership, Ideology, and Identity* (New Haven: Yale University Press, 1989), p. 164. The second El Congreso convention was held in December, 1939 [p. 335, fn. 54].

8. Myra Marx Ferree, "Working Class Feminism: A Consideration of the Consequences of Employment," *The Sociological Quarterly,* Vol. 21 (Spring, 1980): 175. See Chapter 4, "Women and UCAPAWA" in Vicki L. Ruiz, *Cannery Women, Cannery Lives: Mexican Women, Unionization, and the California Food Processing Industry, 1930–1950* (Albuquerque: University of New Mexico Press, 1987).

9. Letter from Luisa Moreno to the author, August 12, 1983.

10. Interview with Dorothy Ray Healey, January 21, 1979, conducted by the author; interview with Carmen Bernal Escobar, February 11, 1979, conducted by the author; interview with Luisa Moreno, September 6, 1979, conducted by the author. See Chapter 4, *Cannery Women, Cannery Lives* for more detail.

11. Interview with Julia Luna Mount, November 17, 1983, conducted by the author; letter from Shifra M. Goldman to the author, June 26, 1984; *Los Angeles Times,* June 12, 1983; García, *Mexican Americans,* p. 220.

12. Cynthia E. Orozco, "Hernández, María L. De," in *Handbook of Texas* (Austin: The Texas State Historical Association, 1996), pp. 572–73; Cotera, *Diosa y Hembra*, pp. 73–82. [Quotes are from Orozco, p. 573, and Cotera, pp. 81–82, respectively.]

13. This phrase is often heard when the man in question appears to have said something or behaved in a way that seemingly belied his feminist principles. Or the phrase can signify a political pedestal on which the compañero is placed.

14. Thomas P. Carter, *Mexican Americans in School: A History of Educational Neglect* (Princeton: College Entrance Examination Board, 1970), p. 31. The seven schools included University of Texas; University of Colorado; University of Arizona; University of California, Riverside; Cal State, Los Angeles; Northern Arizona University; and New Mexico Highlands.

15. Marín, *Social Protest in an Urban Barrio* pp. 114–16; Muñoz, *Youth, Identity, Power*, pp. 58–59, 130.

16. *Chicano Student News*, March 15, 1968; "Proposals Made by High School Students of East Los Angeles to Board of Education, March, 1968," Chicano Blowouts Collection, Chicano Resource Center, East Los Angeles Public Library; Edward J. Escobar, "The Dialectics of Repression: The Los Angeles Police Department and the Chicano Movement, 1968–1971," *Journal of American History*, 79:4 (March 1993): 1495; Script for Episode 3, "The Blowouts: The Struggle for Educational Reform," of *CHICANO!* (Galán Production, 1996), pp. 6–7, 10–18. [Quote is from p. 6.] *Note:* The most detailed overview of events leading to the walkouts is Angel Cervantes's "'The Beginning of Our Being . . .': The East Los Angeles Rebellion of 1968" (M.A. thesis, Claremont Graduate School, 1996).

17. Episode 3, *Chicano!*, p. 5.

18. Rudolfo Acuña, *Community Under Siege: A Chronicle of Chicanos East of the Los Angeles River, 1945–1975* (Los Angeles: UCLA Chicano Research Studies Center, 1984), p. 142; Rudolfo Acuña, *Occupied America*, 2nd ed. (New York: Harper & Row, 1981), p. 358; Episode 3, *Chicano!*, pp. 11–14. [Quote is on p. 14.] *Note:* Sal Castro and twelve others were indicted on conspiracy charges for their participation in the blowouts. The case against the LA Thirteen was thrown out of court two years later.

19. Muñoz, *Youth, Identity, Power*, pp. 64–65; Episode 3, *CHICANO!*, pp. 14, 16.

20. *La Raza*, 1:11 (March 31, 1968): 6; Escobar, "The Dialectics of Repression," pp. 1495–96; *Chicano Student News*, March 15, 1968; *Chicano Student News*, April 25, 1968; Episode 3, *CHICANO!*, pp. 20–21, 35. [Quotes are on pp. 20 and 35, respectively.] Dolores Bernal, a doctoral student in the School of Education at UCLA, is writing her dissertation on Chicana leadership in the blowouts.

21. Acuña, *Occupied America*, p. 358; Muñoz, *Youth, Identity, Power*, pp. 68–70. 131–32, 158. Ana Castillo, *Massacre of the Dreamers: Essays on Xicanisma* (Albuquerque: University of New Mexico Press, 1994), pp. 24, 26. [Quote is on p. 26.]

22. Episode 1, "The Land," of *CHICANO!*, pp. 21–22, 25–26; Muñoz, *Youth, Identity, Power*, pp. 75–78.

23. Muñoz, *Youth, Identity, Power*, p. 78.

24. *Ibid.*, pp. 76–84; Ramón A. Gutiérrez, "Community, Patriarchy, and Individualism: The Politics of Chicano History and the Dream of Equality," *American Quarterly*, 45 (March 1993): 46–47; Juan Gómez-Quiñones, *Mexican Students Por La Raza* (Santa Barbara: Editorial La Causa, 1978), pp. 13, 15. *Note:*

MECHA continues to be the most visible Chicano student group on college campuses.

25. Muñoz, *Youth, Identity, Power,* p. 192.

26. *Ibid.,* pp. 7, 15–16, 51; Gómez-Quiñones, *Mexican Students,* pp. 15–16. For information on textbook stereotypes, see Vicki L. Ruiz, "Teaching Chicano/American History: Goals and Methods," *The History Teacher* 20 (February 1987): 167–68.

27. David G. Gutiérrez, *Walls and Mirrors: Mexican Americans, Mexican immigrants, and the Politics of Ethnicity* (Berkeley: University of California Press, 1995), pp. 184–85, 190–93; Albert Camarillo, *Chicanos in California* (San Francisco: Boyd and Fraser, 1984), pp. 92–103. Carlos Muñoz in *Youth, Identity, Power* offers the most detailed treatment of student participation in La Raza Unida as well as in the genesis of Chicano Studies programs [pp. 113–16; 124–25, 127–67]. The most comprehensive case study of Chicano activism is Ernesto Chávez's "Creating *Aztlán:* The Chicano Student Movement in Los Angeles, 1966–1978" (Ph.D. dissertation, University of California, Los Angeles, 1994).

28. Martínez, "Chingón Politics," p. 47.

29. Mirandé and Enríquez, *La Chicana,* pp. 24–31; Octavio Paz, *Labyrinth of Solitude: Life and Thought in Mexico* (New York: Grove Press, 1961), pp. 65–87; Adelaida R. Del Castillo, "Malintzin Tenépal: A Preliminary Look into a New Perspective," in *Essays on la Mujer,* eds. Rosaura Sánchez and Rosa Martínez Cruz (Los Angeles: University of California Chicano Studies Publications, 1977), pp 124–40; and Ramón Gutiérrez, "Community, Patriarchy, and Individualism," pp. 51–52. *Note:* My family and I were frequent customers of the El Cerezo Restaurant in downtown El Paso during 1982–1985 when I was an assistant professor at U.T. El Paso. I understand that the restaurant is no longer in business.

30. Mirandé and Enríquez, *La Chicana,* pp. 27–28, 247. The original stanzas in Spanish read as follows:

> Se nos quedó el Maleficio
> de brindar al extranjero
> nuestra fe, nuestra cultura,
> nuestro pan, nuestro dinero.
> . . .
>
> Oh! Maldición de Malinche!
> enfermedad del presente,
> cuándo dejarás mi tierra?
> cuándo, háras libre a mi gente!

31. Del Castillo, "Malintzin Tenépal," pp. 124–26, 141–43.

32. Lucha Corpi, "Marina Mother," trans. Catherine Rodríguez-Nieto in *Infinite Divisions,* eds. Rebolledo and Rivero, p. 196.

33. Cordelia Candelaria, "La Malinche, Feminist Prototype," *Frontiers,* 5 (Summer 1980): 6; Carmen Tafolla, "La Malinche," in *Infinite Divisions,* eds. Rebolledo and Rivero, pp. 198–99.

34. Naomi Helena Quiñonez, "Hijas de la Malinche (Malinche's Daughters): The Development of Social Agency Among Mexican American Women and the Emergence of First Wave Chicana Cultural Production "(Ph.D. dissertation, Claremont Graduate School, 1996), pp. 226–27. I thank Naomi for our conversations regarding her experiences as a Chicana cultural worker.

196 Notes to Chapter 5

35. Examples of *El Grito, Encuentro Femenil,* and *Regeneración* can be found in the Colección Tloque Nahuaque, Chicano Studies Library, University of California, Santa Barbara. [Hereinafter referred to as CTN.] A complete run of *El Chicano* is available on microfilm in the Newspaper Files, University Library, University of California, Riverside. For a lively and intriguing examination of literary rebellions against sexism and familial oligarchy among Chicana writers, see Ramón Gutiérrez, "Community, Patriarchy, and Individualism."

36. Anna Nieto Gómez, "La Feminista," *Encuentro Femenil,* 1:2 (1973): 36, CTN; Sonia A. López, "The Role of the Chicana within the Student Movement," in *Essays on La Mujer,* eds. Sánchez and Martínez Cruz, pp. 23–24; Enriqueta Longauex y Vasquez, "The Mexican-American Woman," in *Sisterhood Is Powerful,* ed. Robin Morgan (New York: Vintage Books, 1970), p. 379.

37. López, "The Role of the Chicana," p. 24; Francisca Flores, "Comisión Femenil Mexicana," *Regeneración* 2:1 (1971): 6, CTN.

38. Mirta Vidal, "New Voice of La Raza: Chicanas Speak Out," *International Socialist Review,* 32:9 (October 1971): 7; Elizabeth Olivarez, "Women's Rights and the Mexican American Women," *Regeneración,* 2:4 (1974): 40–42, CTN; López, "The Role of the Chicana," pp. 24–25. [Quotes are from López, *loc. cit.*]

39. Antonio Carmejo, ed. "Why a Chicano Party? Why Chicano Studies?" (pamphlet) (New York: Pathfinder Press, 1970), p. 9.

40. *Ibid.,* p. 10.

41. Adaljiza Sosa Riddell, "Chicanas and El Movimiento," *Aztlán* 5:1 (1974): 156.

42. Gutiérrez, "Community, Patriarchy, and Individualism," p. 49; "El Movimiento and The Chicana," *La Raza* 1:6 (1971): 40; *La Causa,* 1:2 (July 10, 1969), p. 6.

43. *La Causa,* 1:2 (July 10, 1969), *loc. cit.*

44. Nieto Gómez, "La Feminista," pp. 34–37; Jennie V. Chávez, "Women of the Mexican American Movement," *Mademoiselle* (April 1972): 82, 150–52; Adelaida Del Castillo, "La Vision Chicana," *Regeneración,* 2:4 (1974): 48, CTN; Anna Nieto Gómez, "The Chicana Perspective for Education," *Encuentro Femenil,* 1:1 (Spring 1973): 50–53, CTN; "El Movimiento and The Chicana," p. 42. [Quotes are from Del Castillo and "El Movimiento," respectively.] In her early articles, Ana Nieto Gómez's name appears in print as Anna Nieto Gómez.

45. Sylvia Gonzalez, "The White Feminist Movement: The Chicana Perspective," *The Social Science Journal,* 14:2 (April 1977): 70–71; Anna Nieto Gómez, "Chicana Feminism" (undated pamphlet), p. 5, CTN. [Quote is from Nieto Gómez, "Chicana Feminism," *loc. cit.*]

46. Gutiérrez, "Community, Patriarchy and Individualism," p. 66.

47. Del Castillo, "La Vision Chicana," *loc. cit.*

48. Cherríe Moraga and Gloria Anzaldúa, eds., *This Bridge Called My Back,* 2nd ed. (New York: Kitchen Table Press, 1983), see especially pp. 61–101; Chrystos, "I Don't Understand Those Who Have Turned Away from Me," in *Bridge,* eds. Moraga and Anzaldúa, p. 69. For a scholarly overview of the divisions between Chicana and Euro-American feminists, see Alma M. García, "The Development of Chicana Feminist Discourse, 1970–1980," in *Unequal Sisters: A Multicultural Reader in U.S. Women's History,* 2nd ed., eds. Vicki L. Ruiz and Ellen Carol DuBois (New York: Routledge, 1994), pp. 531–44.

49. Jo Carrillo, "And When You Leave, Take Your Pictures with You," in *Bridge,* eds. Moraga and Anzaldúa, pp. 63–64. This exotic othering has persisted.

As I told a reporter for the *Chronicle of Higher Education,* "For the most part Chicanas are still viewed as exotic in women's studies, so we're marginalized." [Karen Winkler, "Scholars Say Issues of Diversity Have Revolutionized Chicano Studies," *Chronicle of Higher Education,* September 26, 1990.]

50. Denise A. Segura and Beatríz M. Pesquera, "Beyond Indifference and Antipathy: The Chicana Movement and Chicana Feminist Discourse," *Aztlán,* 10 (Fall 1988–1990): 70; Quiñonez, "Hijas de la Malinche," pp. 195–97.

51. Nieto Gómez, "La Femenista," pp. 34–35.

52. Sosa Riddell, "Chicanas and El Movimiento," p. 156; Nieto Gómez, "La Femenista," pp. 35–36; Ana Nieto Gómez de Lazarin, "The Chicana Perspective for Education" (ca. 1973), CTN, p. 53; García, "Chicana Feminist Discourse," pp. 536–38.

53. Elizabeth Salas, *Soldaderas in the Mexican Military, Myth and History* (Austin: University of Texas Press, 1990), pp. 115–22; *La Causa,* 1:10 (1970), p. 10; Norma Cantú, "Women, Then and Now: An Analysis of the Adelita Image versus the Chicana as Political Writer and Philosopher," in *Chicana Voices: Intersections of Class, Race, and Gender,* eds. Teresa Córdova, et al. (Austin: UT Austin Center for Mexican Studies Publications, 1986), pp. 8–10. [Quote is from p. 9.]

54. Salas, *Soldaderas,* pp. 115–17, 122; *La Raza* 2 (ca. November–December, 1969), courtesy of Ernesto Chávez.

55. *La Raza* 2 (ca. November–December 1969).

56. *Ibid.*

57. Marta Cotera, "Mexicano Feminism," *Magazín* 1:9 (September, 1973), p. 30; María Eugenia Matute-Bianchi, "A Chicana in Academe" (typescript), pp. 4, 7, CTN; Quiñonez, "Hijas de la Malinche," p. 261; Muñoz, *Youth, Identity, Power,* p. 160, Angie Chabram-Dernesesian, "I Throw Punches for My Race, but I Don't Want to Be a Man: Writing Us-Chica-nos (Girl, Us)/Chicanas —into the Movement Script," *Cultural Studies,* eds. Lawrence Grossberg et al. (New York: Routledge, 1992), p. 83. [Quotes are from Cotera, "Mexicano Feminism," and Chabram-Dernesesian, "I Throw Punches," respectively.] *Note:* One of Naomi Quiñonez's narrators stated that she couldn't say whether or not Chicano professors were involved in the harassment on Nieto Gómez. However, she described how her Chicano assailants "slashed her tires, they broke her windshield . . . [and] somebody threatened her with a gun." Doctoral student at UC Santa Cruz Maylei Blackwell is currently writing about Ana Nieto Gómez and her contributions to Chicana feminist discourse.

58. Sosa Riddell, "Chicanas and El Movimiento," pp. 162–63; Nieto Gómez, "Chicana Feminism," p. 3; Del Castillo, "La Visión Chicana," pp. 1, 3.

59. Acuña, *Occupied America,* p. 403; Virginia Espino, "Women Sterilized as You Give Birth: Forced Sterilization and Chicana Resistance in the 1970s" (paper presented at the National Association for Chicana and Chicano Studies, Chicago, Il: March 21–23, 1996), pp. 8–9. [Quote is from Espino, "Women Sterilized," *loc. cit.*]

60. Espino, "Women Sterilized," pp. 3, 9–10; Flores, "Comisión Femenil Mexicana," pp. 6–8; Carlos G. Velez I., "The nonconsenting sterilization of Mexican women in Los Angeles. Issues of psychocultural rupture and legal redress in paternalistic behavioral environments," in *Twice a Minority: Mexican American Women,* ed. Margarita B. Melville (St. Louis: C.V. Mosby, Co., 1980), pp. 239–40, 244. [Quote is from Espino, "Women Sterilized," pp. 9–10. *Note:* Many Comisión women had been active in the Chicano Student Movement and envi-

sioned their organization as a means of applying their education for community empowerment. For an insider's look at the sterilization suit, see Antonia Hernández "Chicanas and the Issue of Involuntary Sterilization: Reforms Needed to Protect Informed Consent." *Chicano Law Review*, 3:3 (1976): 3–37. [In the footnotes of the article Richard Avila is listed as a co-author.]

61. Hernández, "Chicanas and Sterilization," p. 9. This article provides searing testimony of the impact sterilization had on the lives of the women who filed suit.

62. Velez-I., "The nonconsenting sterilization," p. 246.

63. *Ibid.*; Espino, "Women Sterilized," p. 11; Acuña, *Occupied America*, p. 405. [Quote is from Velez-I., "The nonconsenting sterilization," *loc. cit.*] Examples of these news articles can be found in the *Los Angeles Times*, December 2–3, 1974.

64. Acuña, *Occupied America*, p. 403; Ana Nieto Gómez, "Chicanas in the Labor Force," *Encuentro Femenil*, 1:2 (n.d.), pp. 29–30, CTN; [Quotes are from Acuña, *Occupied America*, *loc. cit.*, Nieto Gómez, "Chicanas in Labor Force," p. 30, respectively.]

65. Yolanda M. Nava, "The Chicana Employment Needs Analysis and Recommendations for Regulation," *Regeneración*, 2:3 (1973), p. 8, CTN.

66. *Regeneración*, 2:3 (1973), p. 3, CTN. For a glimpse at the lobbying efforts of Comisión, see Nava, "The Chicana and Employment," pp. 7–9.

67. Chávez, "Creating Aztlán," pp. 126–42; Escobar, "The Dialectics of Repression," pp. 1505–1506; Ralph Guzmán, "Mexican American Casualties in Vietnam," *La Raza*, 1;1, p. 12 as quoted in Acuña, *Occupied America*, p. 367; Adaljiza Sosa Riddell, "Como Duele," in *Infinite Divisions*, eds. Rebolledo and Rivero, p. 214. *Note:* Because of police and FBI harassment, intimidation, and violence, militant organizations, such as the Chicano Moratorium and the Brown Berets, would disband during the early 1970s. [Chávez, "Creating Aztlán," pp. 139–40; Escobar, "The Dialectics of Repression," p. 1506; Muñoz, *Youth, Identity, Power*, pp. 172–74.]

68. *La Raza*, 1:11 (March 31, 1968), p. 7. Gracias a Angel Cervantes for providing me with this issue.

69. Chávez, "Creating Aztlán," pp. 112–23; Escobar, "The Dialectics of Repression," pp. 1500–1504; Acuña, *Occupied America*, pp. 367–71; Episode One, CHICANO!, p. 34. [The Esparza quote is from Episode One.] According to historian Edward Escobar, the rally became a riot "when the owner of a liquor store a block away from the rally . . . called [police] to complain that teenagers had stolen cold drinks. Sheriff's deputies responded in squad cars with their sirens and lights blazing. The young people who allegedly stole the drinks threw rocks and bottles at the deputies and then ran off toward the . . . demonstration. The deputies followed." [Escobar, "The Dialectics of Repression," pp. 1500–1501.]

70. Episode Four: "Raza Unida" of CHICANO!, pp. 12, 15–19, 36; Cotera, *Diosa y Hembra*, pp. 80, 107–108, 169–70; Carlos Muñoz, *Youth, Identity, Power*, pp. 99–113; Chávez, "Creating Aztlán," pp. 144–62. The only book-length study of La Raza Unida is Ignacio García's *United We Win: The Rise and Fall of La Raza Unida Party* (Tucson: University of Arizona Press, 1989). García, unfortunately, includes little information on women's participation.

71. Cotera, *Diosa y Hembra*, pp. 107–108, 168–73; Cotera, "Mexicano Feminism," *loc. cit.*; Episode Four, CHICANO!, p. 18.; Cynthia E. Orozco, "Mujeres Por La Raza," in *Handbook of Texas*, pp. 878–79; Mary Ann Villarreal, "Tejanas of

La Raza Unida" (graduate seminar paper, Arizona State University, 1996). Luz Gutiérrez further noted that "half the men" left the party rather than share power with women, although, according to Orozco, women formed approximately one-third of the delegates at the party's first state meeting in 1971. [The Gutiérrez quote in the text is from Episode Four, *loc. cit.*]

72. Muñoz, *Youth, Identity, Power,* pp. 113–16, 120–24; Chávez, "Creating Aztlán," pp. 150–55, 178; Carmejo, "Why a Chicano Party?," p. 6. [Quotes are from Chávez, "Creating Aztlán," p. 174, and Carmejo, "Why a Chicano Party?," p. 6, respectively.] California activist Armando Navarro dramatically summarized this nationalist vision: "La Raza Unida must become more than a political organization. It must symbolize the creation of a nation within a nation, a spiritual unification for effective action of all persons of Mexican descent in the United States." [Muñoz, p. *Youth, Identity, Power,* p. 120.]

73. Episode Four, pp. 37–38. Not all Mexican American political leaders would agree with this assessment. Respected Texas Congressman Henry B. González believed that LRU represented "false and mistaken voices of hatred." [*Ibid.*, p. 26.]

74. Mario T. García, *Memories of Chicano History: The Life and Narrative of Bert Corona* (Berkeley: University of California Press, 1994), pp. 297–308; Gutiérrez, *Walls and Mirrors,* pp. 190–92. The section on CASA is also informed by conversations with friends (including my compañero Victor Becerra) who shared with me their insights as insiders.

75. Chávez, "Creating Aztlán," pp. 186–92; García, *Memories of Chicano History,* pp. 297–315. [Quote is from p. 314.]

76. Chávez, "Creating Aztlán," pp. 200–203.

77. Adelaida Del Castillo, ed., *Between Borders: Essays on Mexicana/Chicana History* (Encino: Floricanto Press, 1990), p. i; "Raiz Fuerte que no se Arranca," memorial program in honor of Magdalena Mora (June 13, 1981). Gracias a Devra Weber for sending me a copy. [Quotes are from the memorial program, pp. 17, 23, and 20, respectively.] I also refer readers to her article, "The Role of Mexican Women in the Richmond Tolteca Strike," in *Mexican Immigrant Workers in the U.S.*, ed. Antonio Ríos-Bustamante (Los Angeles: UCLA Chicano Studies Research Center Publications, 1981), pp. 111–18 and the anthology she co-edited with Adelaida Del Castillo, *Mexican Women in the United States* (cf. fn 2).

78. Marisela Chávez, "'We Lived and Breathed and Worked the Movement': Women in el Centro Acción Social Autonoma (CASA), 1975–1978" (graduate seminar paper, Arizona State University, 1996), pp. 3, 11–16. [Quote is on p. 16.] Conversations with former CASA members reveal a gamut of perspectives regarding gender politics within the organization. The work-in progress of Marisela Chávez, a history graduate student at ASU, holds much promise for getting beneath the rhetoric and nostalgia that surrounds CASA.

79. *Ibid.,* pp. 19, 22. [Quote is on p. 22.] Informal conversations with several Casistas suggest that some compañeras, such as Adelaida del Castillo, defined and embraced Chicana feminism and certainly women placed issues, such as involuntary sterilization, on the table of the organization.

80. Velia G. Hancock, "La Chicana, Chicano Movement, and Women's Liberation," *Chicano Studies Newsletter,* University of California, Berkeley (February–March 1971): 6, CTN; Martha P. Cotera, *The Chicana Feminist* (Austin: Information Systems Development, 1977), p. 31.

81. Gutiérrez, *Walls and Mirrors,* p. 196; interview with Graciela M.

Moreno, January 10, 1983, conducted by the author; interview with Ernest Moreno, January 10, 1983, conducted by the author.

82. Muñoz, *Youth, Identity, Power*, p. 201; personal observations by the author.

83. "University Avenue," reprinted from *Borders* by Pat Mora (Houston: Arte Público Press, 1986), p. 19. *Cuentos* are stories.

84. Marcela Christine Lucero Trujillo, "Machismo Is Part of Our Culture," as quoted in Gutiérrez, "Community, Patriarchy, and Individualism," p. 48.

85. Matute-Bianchi, "A Chicana in Academe," p. 3; personal experience of author. In fairness, I want to add that this scholar, with age and maturity, has changed his tune and has demonstrated support for gender issues.

86. Ellen Carol DuBois and Vicki L. Ruiz, eds., *Unequal Sisters: A Multicultural Reader in U.S. Women's History* (New York: Routledge, 1990), p. xv.

87. Cherríe Moraga, "La Guera," in *Bridge*, eds. Anzaldúa and Moraga, pp. 28–29.

88. García, "Chicana Feminist Discourse," p. 537; Martínez, "Chingón Politics," p. 46.

89. María C. González, "Cultural Conflict: Introducing the Queer in Mexican-American Literature Classes," in *Tilting the Tower: Teaching/Lesbians in the Queer Nineties*, ed. Linda Garber (New York: Routledge, 1994), p. 59.

90. Gloria Anzaldúa, *Borderlands/La Frontera: The New Mestiza* (San Francisco: Spinsters/aunt lute, 1987); Gloria Anzaldúa, ed., *Making Face, Making Soul: Haciendo Caras* (San Francisco: an aunt lute foundation book, 1990); Cherríe Moraga, *Loving in the War Years* (Boston: South End Press, 1983); Juanita Ramos, ed., *Compañeras: Latina Lesbians (An Anthology)* (New York: Latina Lesbian History Project, 1987); Carla Trujillo, *Chicana Lesbians: The Girls Our Mothers Warned Us About* (Berkeley: Third Woman Press, 1991). See also Norma Alarcón, Ana Castillo, and Cherríe Moraga, "The Sexuality of Latinas," special issue of *Third Woman*, 4 (1989) and Alicia Gaspar de Alba, "*Tortillerismo*: Work by Chicana Lesbians," *Signs*, 18:4 (Summer 1993): 956–63. Gaspar de Alba's review essay provides an excellent introduction to this literature.

91. Anzaldúa, *Borderlands*, p. 80.

92. Yolanda Levya is a doctoral student in history at the University of Arizona and Alice Hom is a Ph.D. candidate in history at the Claremont Graduate School. In 1996, Ms. Hom received the National Women's Studies Association Dissertation Award for Lesbian Studies. Her dissertation-in-progress is titled "Creating Sites of Resistance: Lesbian of Color Organizing and Activism in Los Angeles and New York, 1969 to the Present."

93. Trujillo, *Chicana Lesbians*, p. ix; Emma Pérez, "Sexuality and Discourse: Notes from a Chicana Survivor," in *Chicana Lesbians*, p. 178; Personal observations by the author who attended the workshop.

94. Carmen Tafolla, *To Split a Human* (San Antonio: Mexican American Cultural Center, 1985), p. 12.

95. Personal experience of the author.

96. *Noticiera de M.A.L.C.S.*, 1 (November 1983): 1.

97. Letter to Professor Ruth Schwarz from MALCS chapter members and concerned students at UCLA dated June 20, 1989 (copy in author's possession). Perhaps as the result of these protests, UCLA hired a Chicana playwright Edith Villarreal the following year for a tenure-track position.

98. *Ibid.*
99. Córdovoa, "Roots of Resistance," p. 175.
100. Castillo, *Massacre of the Dreamers;* Anzaldúa, *Making Face, Making Soul,* p. xxvii.
101. Segura and Pesquera, "Beyond Indifference," pp. 69–88. [Quotes taken from pp. 81, 85, respectively.]
102. Segura and Pesquera, "Beyond Indifference and Antipathy," p. 82.
103. Cynthia Orozco, "Sexism in Chicano Studies and the Community," in *Chicana Voices,* p. 13; Anzaldúa, "Foreword to the Second Edition," *Bridge,* p. iv.

Chapter 6

1. Interview with Elsa Chávez, April 19, 1983, conducted by the author; Laurie Coyle, Gail Hershatter, and Emily Honig, "Women at Farah: An Unfinished Story," in *Mexican Women in the United States: Struggles Past and Present* (Los Angeles: UCLA Chicano Studies Research Center Publications, 1980), pp. 117, 119; The San Francisco Bay Area Farah Strike Support Committee, "Chicanos Strike at Farah," pamphlet (January, 1974), pp. 2–3, 6–7; letter from Sidney Metzger, Bishop of El Paso to the Bishop of Rochester dated October 31, 1972 ,reprinted in *Viva La Huelga: Farah Strike Bulletin no. 15* (ACW newsletter) *Note:* Elsa Chávez is a pseudonym used at the person's request.
2. Coyle, Hershatter, and Honig, "Women at Farah," p. 135.
3. Allen Pusey, "Clothes Made the Man," *Texas Monthly* (June, 1977), pp. 133, 136–37; "A boycott to aid garment workers," *Business Week,* August 26, 1972: 53–54; *Lubbock Avalanche Journal,* November 26, 1972; *El Paso Times,* June 6, 1983; SF Farah Strike Support Committee, "Chicanos Strike at Farah," p. 9; John Rebchook, "El Paso Is a Minimum Wage Town," *Special Report: The Border* (*El Paso Herald Post,* Summer 1983), p. 74. *Note:* In 1983, the El Paso Chamber of Commerce honored Willie Farah as its "Manufacturer of the Year," an "award bestowed on a businessman who has excelled in the advancement of his career while demonstrating an exceptional measure of concern for the community" (*El Paso Times,* June 6, 1983).
4. SF Farah Strike Support Committee, "Chicanos Strike at Farah," pp. 4, 7; Coyle, Hershatter, and Honig, "Women at Farah," pp. 127, 128–30; *Viva La Huelga: Farah Strike Bulletin, no. 1; Washington Post,* September 28, 1972; *Business Week,* August 26, 1972.
5. *Viva La Huelga: Farah Strike Bulletin, no. 6; Washington Post,* September 28, 1972.
6. *El Paso Herald Post,* June 27, 1973; *El Paso Times,* November 29, 1973; *El Paso Times,* February 22, 1974; *El Paso Herald Post,* January 11, 1974; *El Paso Herald Post,* January 7, 1974; Paul Newton Poling, "For the Defense of the Farah Workers" (pamphlet published by Farah Manufacturing, ca. 1973); Chávez interview; Coyle, Hershatter, and Honig, pp. 128–30. "Farah Workers Group Diverts Yule Donation from Church" represents a typical headline in El Paso papers (*El Paso Herald Post,* December 22, 1972).
7. *Guardian,* March 6, 1974.
8. Examples of such letters can be found in the *El Paso Herald Post,* November 23, 1973; *El Paso Times,* December 15, 1973; *El Paso Times,* January 14,

1974; *El Paso Herald Post,* January 28, 1974; *El Paso Herald Post,* February 15, 1974; *El Paso Times,* January 25, 1974. Quote is taken from the *El Paso Herald Post,* December 15, 1973.

9. Examples of letters supporting the strikers can be found in the *El Paso Times,* January 14, 1974; *El Paso Times,* January 27, 1974; *El Paso Times,* February 2, 1974. [Quote is taken from the *El Paso Herald Post,* January 23, 1974.]

10. *The Washington Post,* September 28, 1972.

11. Ibid.; *The Advance* [ACW National Newspaper] (January 1973), pp. 5–16; Deborah DeWitt Malley, "How the Union Beat Willie Farah," *Fortune* (August, 1974): 167, 238; *Viva La Huelga, Farah Strike Bulletin, no. 1* (1972); *Viva La Huelga, Farah Strike Bulletin, no. 4* (1972); *Viva La Huelga: Farah Strike Bulletin, no. 8* (1972); *Viva La Huelga, Farah Strike Bulletin, no. 10* (1972); *El Paso Times,* September 12, 1972; *El Paso Herald Post,* February 2, 1973. Note: Senator Gaylord Nelson (D-Wisconsin) chaired the Citizens' Committee for Justice for Farah Workers and the legendary African American labor leader A. Philip Randolph served as honorary chair. National figures who publicly joined the committee included Senator Edward Kennedy, Carey McWilliams, Joanne Woodward, Vernon Jordan, Archibald MacLeish, Ramsey Clark, Clark Kerr, and Arthur Schlesinger, Jr. [*Viva La Huelga, Farah Strike Bulletin, no 4.* (1972). Furthermore, the *El Paso Herald Post,* in announcing the arrival of Democratic vice-presidential candidate Sergeant Shriver, mentions Shriver's "advance man" Bill Clinton. [*El Paso Herald Post,* September 12, 1972. The headline reads, "Democrats Plan Big Love-In."]

12. *Texas Monthly* (June, 1977): 137.

13. Coyle, Hershatter, and Honig, "Women at Farah," pp. 134–35; 141. [Both quotes are taken from p. 134.]

14. Chávez interview. In 1972, Farah employed from 9,000 to 10,000 workers. Farah strike support publications and the classic scholarly study "Women at Farah" place the number of strikers at 4,000. The anti-union piece "In Defense of the Farah Workers" and an article in *Fortune* magazine underestimates their numbers by one-half or 2,000 strikers. The *Washington Post* judges the strikers at 3,000 out of 8,500 workers. [See Coyle, Hershatter, and Honig, "Women at Farah," p. 117; SF Farah Strike Support Committee, "Chicanos Strike at Farah," p. 1; Poling, "In Defense of the Farah Workers," pp. 5, 13; *Fortune* magazine (August 1974): 165, 167; *Washington Post,* September 28, 1972.

15. Chávez interview.

16. Coyle, Hershatter, and Honig, "Women at Farah," pp. 117, 129, 131–34; Chávez interview; Portland [Oregon] Strike Support Committee, "The Farah Strike: Our Struggle, Too," pamphlet (1973), pp. 7, 12.

17. Rose Pesotta, *Bread upon the Waters* (New York: Dodd, Mead, and Company, 1944), pp. 50–51. Note: During the 1959–1961 Tex-Son garment strike in San Antonio, Mexican-American children passed out balloons to children entering local department stores imprinted with the message "Don't Buy Tex-Son Children's Clothes." [Tex-Son strike photographs, George Lambert Collection, Labor Archives, University of Texas, Arlington, Arlington, Texas.] For more information on these strikes, see Clementina Durón, "Mexican Women and Labor Conflict in Los Angeles: The ILGWU Dressmakers Strike of 1933," *Aztlán,* 15:1 (1984): 145–61; Irene Ledesma, "Texas Newspapers and Chicana Workers' Activism, 1919–1972," *Western Historical Quarterly,* 26:3 (Autumn 1995): 309–31.

18. Coyle, Hershatter, and Honig, "Women at Farah," pp. 133–34.

19. SF Farah Strike Support Committee, "Chicanos Strike at Farah," p. 16.

20. Chávez interview; Coyle, Hershatter, and Honig, "Women at Farah," pp. 117, 138–42.

21. Coyle, Hershatter, and Honig, "Women at Farah," p. 143,

22. Chávez interview.

23. Jacques Levy, *Cesar Chávez: Autobiography of La Causa* (New York: W. W. Norton, 1975) provides a comprehensive and intimate portrait of Cesar Chávez and the birth of the United Farm Workers. See especially Book IV, "The Birth of Union," pp. 149–218, and Book V, "Victory in the Vineyards," pp. 219–325. An outstanding scholarly monograph on the UFW is Linda C. Majka and Theo J. Majka, *Farm Workers, Agribusiness, and the State* (Philadelphia: Temple University Press, 1982). Figures on union membership taken from the *Los Angeles Times,* March 27, 1994.

24. Barbara Baer and Glenna Matthews, "Women of the Boycott," in *America's Working Women: A Documentary History–1600 to the Present,* eds. Rosalyn Baxandall and Susan Reverby (New York: Vintage Books, 1976), pp. 363–72; interview with Graciela Martínez Moreno, January 10, 1983, conducted by the author; interview with Ernest Moreno, January 10, 1983, conducted by the author; interview with María del Carmen Romero, January 8, 1983; quote is taken from Levy, *Cesar Chávez,* p. 160. The best historical survey on women's involvement in the United Farm Workers is Margaret Eleanor Rose's "Women in the United Farm Workers: A Study of Chicana and Mexicana Participation in a Labor Union, 1950 to 1980" (Ph.D. dissertation, University of California, Los Angeles, 1988). For further details on family involvement, see Chapter 6 of this dissertation.

25. Martínez Moreno interview; Moreno interview; Romero interview.

26. Martínez Moreno interview; Rose, "Women in the United Farm Workers," pp. 9, 150–51, 243–89.

27. Fran Leeper Buss, ed., *Forged under the Sun/Forjada bajo del Sol: The Life of María Elena Lucas* (Ann Arbor: University of Michigan Press, 1993), pp. 195, 216–17. [Quotes taken from pp. 216 and 217, respectively.]

28. Buss, *Forged under the Sun,* pp. 220–21; 225–31, 258. [Quote is from p. 258.]

29. With regard to sexism within the union, María Elena Lucas stated, "There's been times as a woman with the FLOC that I've felt swatted like a fly, and as a woman, I think that's wrong." [Buss, *Forged under the Sun,* p. 258]. In a recent scholarly study of the Farm Labor Organizing Committee, the authors make no mention of the contributions of María Elena Lucas, referring to her briefly as a victim of pesticide poisoning. "Usually a bright and cheerful woman, she told her story of pesticide poisoning . . . with tears on her face." It is a strictly male-centered narrative that erases women's roles as leaders and rank-and file-activists. See W. K. Barger and Ernesto M. Reza, *The Farm Labor Movement in the Midwest* (Austin: University of Texas Press, 1994). [Quote is from p. 31.]

30. Rose, "Women in the United Farm Workers," pp. x, 7–10, 46, 68–80, 96–103. Quote is taken from p. 101. An excellent portrait of Dolores Huerta can be seen in "Act One: Work and Family," in the TBS series "A Century of Women" (1994). This segment was produced by noted Chicana filmmaker Sylvia Morales.

31. Rudolfo Acuña, *Occupied America : A History of Chicanos,* 3rd ed. (New York: HarperCollins, 1988), p. 269; Alan Covey, ed., *A Century of Women* (At-

lanta: TBS Books, 1994), p. 46. [This is the companion volume to the TBS series of the same name, cf. n. 30.]

32. Covey, *A Century of Women*, p. 47.

33. Levy, *Cesar Chávez*, p. 168.

34. *Ibid.*, pp. 168–69.

35. Rose, "Women in the United Farm Workers," pp. 117, 134–37, 140, 144–51. Quote is from p. 150. Rose refers to "political familialism" in fn 45, p. 160. See Maxine Baca Zinn, "Political Familialism: Toward Sex Role Equality in Chicano Families," *Aztlán*, 8:1 (1975): 13–26.

36. Barbara Kingsolver, *Holding the Line: Women in the Great Arizona Mining Strike of 1983* (Ithaca: ILR Press, 1989), pp. 1–21. [Quote is taken from p. 17.] For historical background on Arizona mining, see James W. Byrkit, *Forging the Copper Collar* (Tucson: University of Arizona Press, 1982) and "Los Mineros," a PBS American Experience series, produced by Hector Galán (1991). *Note:* The historical consultant for "Los Mineros," Christine Marín, a doctoral student and archivist at Arizona State University, is writing on the history of Mexican mining families in Arizona.

37. *Ibid.*, pp. 15, 31–33, 146–49, 157–62, 176–90. [Quotes are from pp. 15, 186, respectively.]

38. Karen Anderson, *Changing Woman: A History of Racial Ethnic Women in Modern America* (New York: Oxford University Press, 1996), pp. 147–48; Acuña, *Occupied America*, p. 448. [Quotes are from Anderson, *Changing Woman*, loc. cit.]

39. Marta P. Cotera, *Diosa y Hembra: The History and Heritage of Chicanas in the U.S.* (Austin: Information Systems Development, 1976) , p. 179.

40. Patricia Zavella, *Women's Work and Chicano Families: Cannery Workers of the Santa Clara Valley* (Ithaca: Cornell University Press, 1987).

41. Louise Lamphere, Patricia Zavella, Felipe Gonzales with Peter B. Evans, *Sunbelt Working Mothers: Reconciling Family and Factory* (Ithaca: Cornell University Press, 1993), p. 285. *Note:* "Hispana" is a common term of self-identification for Mexican-American women in New Mexico and Colorado.

42. Lamphere, Zavella, et al., *Sunbelt Working Mothers*, pp. 290–91. Lamphere's Chapter 5, "Management Ideology and Practice in Participatory Plants" (pp. 138–82), seems particularly enlightening. For more examples of the structural difficulties in organizing women in southwestern industries, see Vicki L. Ruiz, "'And Miles to go . . .': Mexican Women and Work, 1930–1985," in *Western Women, Their Land, Their Lives*, eds. Lillian Schlissel, Vicki L. Ruiz, and Janice Monk (Albuquerque: University of New Mexico Press, 1988): 117–36.

43. *Los Angeles Times*, September 6, 1993; *Los Angeles Times*, August 12, 1992; *Los Angeles Times*, March 20, 1994.

44. *Los Angeles Times*, September 27, 1992.

45. *Los Angeles Times*, March 20, 1994.

46. Mary Helen Ponce, *Hoyt Street: An Autobiography* (Albuquerque: University of New Mexico Press, 1993); Personal observations by the author; Yolanda Tarango, "The Hispanic Woman and Her Role in the Church," *New Theology Review*, Vol. 3 (November 1990): 58. *Cascarones* are eggshells filled with confetti.

47. Joseph D. Sekul, "Communities Organized for Public Service: Citizen Power and Public Policy in San Antonio," in *The Politics of San Antonio: Community, Progress, and Power*, eds. David R. Johnson, John A. Booth, and Richard J. Harris (Lincoln: University of Nebraska Press, 1983), pp. 176–77; Sara M.

Evans, *Born for Liberty: A History of Women in America* (New York: The Free Press, 1989), p. 309.

48. Peter Skerry, *Mexican Americans: The Ambivalent Minority* (New York: The Free Press, 1993), pp. 177–78.

49. Sekul, "Communities Organized for Public Service," pp. 175–90; Skerry, *Mexican Americans,* pp. 176–80; Sidney Plotkin, "Democratic Change in the Urban Political Economy: San Antonio's Edwards Aquifer Controversy," in *The Politics of San Antonio,* pp. 167–71; Acuña, *Occupied America,* pp. 433–34; John A. Booth, "Political Change in San Antonio, 1970–82: Toward Decay or Democracy," in *The Politics of San Antonio,* pp. 195–96, 200. [Quotes are from Sekul, "Communities Organized for Public Service," pp. 175, 181, respectively.]

50. "Adelante Mujeres," video produced by the National Women's History Project (1992); Arnoldo De León, *Mexican Americans in Texas: A Brief History* (Arlington Heights, Ill.: Harlan Davidson, Inc., 1993), pp. 37, 139; Booth, "Political Change in San Antonio," pp. 200, 202, 204.

51. Skerry, *Mexican Americans,* pp. 176–80; Acuña, *Occupied America,* p. 436; Sekul, "Communities Organized for Public Service," p. 190.

52. Evans, *Born for Liberty,* p. 309.

53. Skerry, *Mexican Americans,* p. 149.

54. Evans, *Born for Liberty,* p. 309.

55. *Ibid.,* p. 313.

56. For differing views of contemporary Alinksy-based groups in Los Angeles, contrast Acuña, *Occupied America* with Skerry, *Mexican Americans.*

57. *Los Angeles Times,* May 2, 1994.

58. *Ibid.*

59. Gilbert Cadena and Lara Medina, "Liberation Theology and Social Change: Chicanas and Chicanos in the Catholic Church" (unpublished manuscript), pp. 12–17; Lara Medina, "Calpulli: A Chicano Self-Help Organization," *La Gente* (1990). Gracias a Lara Medina for this news clipping and for our conversations on Latino liberation theology. [Quote is taken from *La Gente.*]

60. Cadena and Medina, "Liberation Theology and Social Change," p. 13.

61. Cadena and Medina, pp. 14–16. I gave an oral history workshop for calpulli in January 1994. During the workshop, Sister Rosa Martha Zarate articulated the need to make links with local Native Americans and recognize the significance of Mesoamerican indigenous traditions.

62. *La Gente* (1990).

63. Mary Pardo, "Identity and Resistance: Mexican American Women and Grassroots Activism in Two Los Angeles Communities" (Ph.D. dissertation, University of California, Los Angeles, 1990); Acuña, *Occupied America,* p. 424; *Los Angeles Herald Examiner,* September 28, 1986. Gracias a Howard Shorr for sending me this clipping. [Quote is from the *Los Angeles Herald Examiner.*]

64. Tom Chorneau, "Molina Lays Siege to Prison," *Los Angeles Downtown News* (ca. June 1986), p. 6. [News clipping courtesy of Howard Shorr.]

65. Skerry, *Mexican Americans,* p. 181; Pardo, "Identity and Resistance"; *Los Angeles Herald Examiner,* September 28, 1986.

66. *Chicago Tribune,* March 20, 1995; *Los Angeles Times,* September 7, 1995; Nina Schuyler, "LA Moms Fight Back," *Progressive,* 56:8 (August 1992): 13. [Quote is from Schuyler, "LA Moms," *loc. cit.*] Gracias a Betsy Jameson for sharing these clippings with me.

67. Richard Steven Street, *Organizing for Our Lives: New Voices from Rural*

Communities (Portland, Ore.: Newsage Press and California Rural Legal Assistance, 1992). Photo is on p. 29. Gracias a Richard Street for sending me a copy of this important book.

68. Street, *Organizing for Our Lives*, pp. 66–81. [Quote is taken from pp. 69–70.]

69. *Ibid.*, p. 68.

70. *Ibid.*, p. 75.

71. *Chronicle of Philanthropy*, July 24, 1990; *New York Times*, August 20, 1990. [Quotes are from the *Chronicle of Philanthropy* and the *New York Times*, respectively.] Gracias a the National Women's History Project for sending me these news clippings. With regard to Los Ganados del Valle's efforts to secure cooperative grazing rights, see Laura Pulido, *Environmentalism and Economic Justice: Two Chicano Struggles in the Southwest* (Tucson: University of Arizona Press, 1996).

72. Eudora Welty, *One Writer's Beginnings* (Cambridge: Harvard University Press, 1983), p. 102.

73. Chela Sandoval, "U.S. Third World Feminism: The Theory and Method of Oppositional Consciousness in the Postmodern World," *Genders* Vol. 4 (1991): 15.

74. *El Paso Times*, February 6, 1974.

Epilogue

1. Anne Larson, "A Comparative Analysis of Proposition 187 and the Immigration Action of 1924" (seminar paper, the Claremont Graduate School, 1994); "U.S. Citizen Opinion Poll on America's Illegal Alien Crisis" by American Immigration Control, a project of "We the People" (1995); *Los Angeles Times*, October 30, 1994; *Los Angeles Times*, March 12, 1995. Proposition 187 purports to deny all public services, except for emergency health care, to undocumented workers and requires doctors and teachers to report to the INS people they suspect as undocumented.

2. Robert Frost, "Stopping by Woods on a Snowy Evening," in *Complete Poems of Robert Frost* (New York: Holt, Rinehart,and Winston, 1958), 275.

3. U.S. Bureau of the Census, *We the American . . . Hispanics* (November 1993), pp. 2–3; According to the 1990 census, the five southwestern states account for over 60 percent of the Latino population in the United States.

4. Denise A. Segura, "Chicanas and Education: Implications for Occupational Mobility" (unpublished paper, 1995).

5. *Ibid.*, p. 9. For an excellent study on Mexican women and aging, see Elisa Facio, *Understanding Older Chicanas* (Thousand Oaks, Cal.: Sage Publications, 1995) and Mexican families, in general, see Norma Williams, *The Mexican American Family: Tradition and Change* (Boston: G.K. Hall, 1990).

6. U.S., Bureau of the Census, *1990 Census of Population. Social and Economic Characteristics: Colorado, Arizona, California, New Mexico, and Texas* (Washington, D.C.: G.P.O., 1993), Tables 50 and 124.

7. Patricia Zavella, *Women's Work and Chicano Families: Cannery Workers of the Santa Clara Valley* (Ithaca: Cornell University Press, 1987), pp. 162–66 [quote is from p. 164]; Vicki L. Ruiz, "'And Miles to Go: Mexican Women and

Work, 1939–1985," in *Western Women: Their Land, Their Lives*, eds., Lillian Schlissel, Vicki L. Ruiz, and Janice Monk (Albuquerque: University of New Mexico Press, 1988), pp. 127–34.

8. Elizabeth Martínez, "Levi's, Button Your Fly—Your Greed Is Showing," *Z Magazine* (January 1993): 22–27. [Quotes are from pp. 24, 26, respectively.]. Martínez points out that "Levi's has closed . . . 26 plants across the U.S. since 1985" (p. 27).

9. Mark Thompson, "Threadbare Justice," *California Lawyer*, Vol. 10 (May 1990): 30, 32, 84; María Angelina Soldatenko, "Organizing Latina Garment Workers in Los Angeles," *Aztlán*, Vol. 20:1–2 (1991): 79–81; *Los Angeles Times*, March 12, 1995.

10. Soldatenko, "Organizing Latina Workers," p. 86; *Los Angeles Times*, April 15, 1994; *California Lawyer*, pp. 28–32. [Quote is on pp. 30–31.] According to *California Lawyer*, apparel companies receive "a virtual slap on the wrist" for violating child labor laws. For an insider's view of life as a *costurera*, see María Angelina Soldatenko, "The Everyday Lives of Latina Garment Workers in Los Angeles: The Convergence of Gender, Race, Class, and Immigration" (Ph.D. dissertation, University of California, Los Angeles, 1992).

11. Interview with Angela Barcena, August 2, 1979, conducted by Oscar Martínez, Mario Galdos, and Virgilio Sánchez (on File at the Institute of Oral History (IOH), University of Texas at El Paso). Barcena interview. The English translation is as follows:

> The majority of women in the factories, the majority need the work to help their husbands support the family or they are heads of households. They are the ones that support the families, they are single, they are divorced. That is why the Mexican women has been very susceptible to exploitation.

Barcena's observations, which she made in 1979, continue to the present. On June 7, 1996, an article appearing in the *Los Angeles Times* elaborated on the material conditions of undocumented women wage earners. "Women are less likely than men to speak English and therefore, more at risk for exploitation, more often working below minimum wage, and more often the subjects of sexual harassment and deportation." The journalist, however, cannot attribute exploitation to language barriers alone. As the appended tables indicate, persistent disparities in income and education along gender and racial/ethnic lines belie any single cause or remedy. [*Los Angeles Times*, June 7, 1996. Gracias a Valerie Matsumoto for sending me clippings from the *Los Angeles Times*.]

12. *Los Angeles Times*, February 18, 1996.

13. *Ibid.; Los Angeles Times*, June 7, 1996.

14. "The Confession," reprinted from *Sueño de Colibri/Hummingbird Dream* by Naomi Quiñonez (Los Angeles: West End Press, 1985), p. 57.

15. This speech has recently been published in its entirety. See Luisa Moreno, "Caravans of Sorrow: Noncitizen Americans of the Southwest," in *Between Two Worlds: Mexican Immigrants in the United States*, ed. David G. Gutiérrez (Wilmington, DE: Scholarly Resources, 1996): 119–23.

16. *Sí Magazine*, (Summer 1996): 6. *Note:* contrary to the stereotype of immigrants as a drain on social services, they "are considerably less likely than the native born to receive public assistance." Separate studies by economists and demographers "concluded that public-resource utilization by Mexican undocu-

mented immigrants is negligible, that only 2 to 6 percent have ever received any type of AFDC, Social Security payments, or food stamps." [Pierrete Hondagneu-Sotelo, *Gendered Transitions: Mexican Experiences of Immigration* (Berkeley: University of California Press, 1994), pp. 170–71].

17. Flyer announcing a march against Proposition 187 scheduled for October 16 in Los Angeles (in author's possession); *Los Angeles Times,* October 17, 1994; *La Opinion,* October 17, 1994; Martínez, "Levi's," p. 23.

Bibliography

❖❖❖

Government Documents

California. Mexican Fact-Finding Committee. *Mexicans in California: Report of Governor C.C. Young's Mexican Fact-Finding Committee.* San Francisco: California State Printing Office, 1930; San Francisco: R and E Research Associates, 1970.

U.S. Bureau of the Census. *1990 Census of Population. Social and Economic Characteristics: Arizona.* Washington, D.C.: G.P.O., 1993

———. *Social and Economic Characteristics: California.* Washington, D.C.: G.P.O., 1993

———. *Social and Economic Characteristics: Colorado.* Washington, D.C.: G.P.O., 1983.

———. *Social and Economic Characteristics: New Mexico.* Washington, D.C.: G.P.O., 1983.

———. *Social and Economic Characteristics: Texas.* Washington, D.C.: G.P.O., 1983

———. *We . . . the American Hispanics.* Washington, D.C.: G.P.O., 1993

U.S. Congress. House Committee on Immigration and Naturalization. *Western Hemisphere Immigration: H.R. 8523, H.R. 8530, H.R. 8702.* 72nd Cong., 2nd sess., 1930.

U.S. Congress. Senate Committee on Education and Labor. *Violations of Free Speech and Rights of Labor: Hearings Before a Subcommittee of the Senate Committee on Education and Labor, United States Senate Part 51.* 76th Cong., 2nd sess., 1940.

U.S. Congress. Senate Committee on Education and Labor. *Violations of Free Speech and Rights of Labor: Hearings Before a Subcommittee of the Senate Committee on Education and Labor, United States Senate Part 70.* 76th Cong., 3rd. sess., 1941.

U.S. Congress. Senate Committee on Education and Labor. *Violations of Free Speech and Rights of Labor: Hearings Before a Subcommittee of the Senate Committee on Education and Labor, United States Senate Part 71.* 76th Cong., 3rd. sess., 1941.

U.S. Immigration and Naturalization Service. Records, ser. A: Subject correspondence files, pt. 2. *Mexican Immigration, 1906–1930.* Research Collections in American Immigration. Bethesda, Md.: University Publications of America, 1992. Microfilm.

Books

Acuña, Rudolfo F. *A Community under Siege: A Chronicle of Chicanos East of the Los Angeles River, 1945–1975.* Los Angeles: UCLA Chicano Studies Research Center Publications, 1984.

———. *Occupied America: A History of Chicanos.* 2nd ed. New York: Harper & Row, 1981.

Alarcón, Norma, Ana Castillo, and Cherríe Moraga. "The Sexuality of Latinas." [special issue] *Third Woman* 4 (1989).

Almaguer, Tomás. *Racial Fault Lines: The Historical Origins of White Supremacy in California.* Berkeley: University of California Press, 1994.

Alvarez Jr., Robert R. *Familia: Migration and Adaptation in Baja and Alta California, 1800–1975.* Berkeley: University of California Press, 1987.

Anderson, Karen. *Changing Woman: A History of Racial Ethnic Women in Modern America.* New York: Oxford University Press, 1996.

Anzaldúa, Gloria. *Borderlands/La Frontera: The New Mestiza.* San Francisco: Spinsters/aunt lute, 1987.

———.,ed. *Making Face, Making Soul: Haciendo Caras.* San Francisco: aunt lute foundation, 1990.

Babbitt, C. S. *The Remedy for the Decadence of the Latin Race.* El Paso, Tex.: El Paso Printing Co., ca 1909.

Balderrama, Francisco. *In Defense of La Raza: The Los Angeles Mexican Consulate and the Mexican Community, 1929–1936.* Tucson: University of Arizona Press, 1982.

Balderrama, Francisco, and Raymond Rodríguez. *Decade of Betrayal: Mexican Repatriation in the 1930s.* Albuquerque: University of New Mexico Press, 1995.

Bancroft, Hubert Howe. *California Pastoral, 1769–1848.* The Works of Hubert Howe Bancroft, v. 34. San Francisco: The History Company, 1888.

Barger, W. K., and Ernesto M. Reza. *The Farm Labor Movement in the Midwest.* Austin: University of Texas Press, 1994.

Barrera, Mario. *Race and Class in the Southwest.* Notre Dame, Ind.: University of Notre Dame Press, 1979.

Benhabib, Seyla, and Drucilla Cornell, eds. *Feminism as Critique.* Minneapolis: University of Minnesota Press, 1987.

Biberman, Herbert *Salt of the Earth: The Story of a Film.* Boston: Beacon Press, 1965.

Blackwelder, Julia Kirk. *Women of the Depression: Caste and Culture in San Antonio.* College Station: Texas A&M Press, 1984.

Bogardus, Emory S. *The Mexican in the United States.* Los Angeles: University of Southern California Press, 1934.

Byrkit, James W. *Forging the Copper Collar.* Tucson: University of Arizona Press, 1982.

Cabeza de Baca, Fabiola. *We Fed Them Cactus.* Albuquerque: University of New Mexico Press, 1954.

Camarillo, Albert. *Chicanos in a Changing Society: From Mexican Pueblos to American Barrios in Santa Barbara and Southern California, 1848–1930.* Cambridge: Harvard University Press, 1979.

———. *Chicanos in California.* San Francisco: Boyd and Fraser, 1984.

Cardoso, Lawrence A. *Mexican Emigration to the United States, 1897–1931: Socio-Economic Patterns.* Tucson: University of Arizona Press, 1980.

Carter, Thomas P. *Mexican Americans in School: A History of Educational Neglect.* Princeton: College Entrance Examination Board, 1970.

Castillo, Ana. *Massacre of the Dreamers: Essays on Xicanisma.* Albuquerque: University of New Mexico Press, 1994.

Chafe, William H. *The American Woman: Her Changing Social, Economic, and Political Roles, 1920–1970.* New York: Oxford University Press, 1972.

Clark, Margaret. *Health in the Mexican American Culture.* Berkeley: University of California Press, 1959.

Cohen, Lizbeth. *Making a New Deal: Industrial Workers in Chicago, 1919–1939.* Cambridge: Cambridge University Press, 1990.

Cortera, Martha. *Diosa y Hembra: The History and Heritage of Chicanas in the U.S.* Austin, Tex.: Information Systems Development, 1976.

Cotera, Martha P. *The Chicana Feminist.* Austin, Tex.: Information Systems Development, 1977.

Covey, Alan, ed. *A Century of Women.* Atlanta: TBS Books, 1994.

Daniel, Cletus E. *Bitter Harvest: A History of California Farmworkers, 1870–1941.* Ithaca: Cornell University Press, 1981.

de la Torre, Adela, and Beatríz M. Pesquera, eds. *Building With Our Hands: New Directions in Chicana Studies.* Berkeley: University of California Press, 1993.

Del Castillo, Adelaida R., ed. *Between Borders: Essays on Mexicana/Chicana History.* Encino, Calif.: Floricanto Press, 1990.

De León, Arnoldo. *Mexican Americans in Texas: A Brief History.* Arlington Heights, Ill.: Harlan Davidson, 1993.

D'Emilio, John, and Estelle B. Freedman. *Intimate Matters: A History of Sexuality in America.* New York: Harper & Row, 1988.

Deutsch, Sarah. *No Separate Refuge: Culture, Class, and Gender on the Anglo-Hispanic Frontier in the American Southwest, 1880–1940.* New York: Oxford University Press, 1987.

DuBois, Ellen Carol, and Vicki L. Ruiz, eds. *Unequal Sisters: A Multicultural Reader in U.S. Women's History.* New York: Routledge, 1990.

Ellis, Pearl Idella. *Americanization through Homemaking.* Los Angeles: Wetzel, 1929.

Evans, Sara M. *Born for Liberty: A History of Women in America.* New York: The Free Press, 1989.

———. *Personal Politics: The Roots of Women's Liberation in the Civil Rights Movement and the New Left.* New York: Vintage, 1980.

Ewen, Elizabeth. *Immigrant Women in the Land of Dollars.* New York: Monthly Review Press, 1985.

Ewen, Stuart, and Elizabeth Ewen. *Channels of Desire.* New York: McGraw-Hill, 1982.

Facio, Elisa. *Understanding Older Chicanas.* Thousand Oaks, CA: Sage Publications, 1995.

Fass, Paula. *The Damned and the Beautiful.* New York: Oxford University Press, 1977.

Fitzgerald, F. Scott. *Flappers and Philosophers.* London: W. Collins Sons, 1922.

Fraser, Nancy. *Unruly Practices: Power, Discourse, and Gender in Contemporary Social Theory.* Minneapolis: University of Minnesota Press, 1989.

Gabaccia, Donna R. *From Sicily to Elizabeth Street: Housing and Social Change among Italian Immigrants, 1880–1930.* Albany: State University of New York Press, 1984.

Gamio, Manuel. *The Life Story of the Mexican Immigrant.* Chicago: University of Chicago Press, 1931; New York: Dover Publications, 1971.

———. *Mexican Immigration to the United States.* Chicago: University of Chicago Press, 1930; New York: Arno Press, 1969.

García, Ignacio. *United We Win: The Rise and Fall of La Raza Unida Party.* Tucson: University of Arizona Press, 1989.

García, Mario T. *Desert Immigrants: The Mexicans of El Paso, 1880–1920.* New Haven: Yale University Press, 1981.

———. *Memories of Chicano History: The Life and Narrative of Bert Corona.* Berkeley: University of California Press, 1994.

———. *Mexican Americans: Leadership, Ideology, and Identity, 1930–1960.* New Haven: Yale University Press, 1989.

García, Richard A. *Rise of the Mexican American Middle Class: San Antonio, 1929–1941.* College Station: Texas A&M Press, 1991.

Glenn, Susan A. *Daughters of the Shtetl.* Ithaca: Cornell University Press, 1990.

Gluck, Sherna Berger. *Rosie the Riveter Revisited: Women, the War and Social Change.* Boston: Twayne, 1987.

Gómez-Quiñones, Juan. *Mexican Students Por La Raza.* Santa Barbara: Editorial La Causa, 1978.

———. *Roots of Chicano Politics, 1600–1940.* Albuquerque: University of New Mexico Press, 1994.

González, Deena J. *Refusing the Favor: Spanish-Mexican Women of Santa Fé, 1820–1880.* New York: Oxford University Press, forthcoming.

González, Gilbert. *Chicano Education in the Era of Segregation.* Philadelphia: Balch Institute Press, 1990.

———. *Labor and Community: Mexican Citrus Worker Villages in a Southern California County, 1900–1950.* Urbana: University of Illinois Press, 1994.

Griswold del Castillo, Richard. *La Familia: Chicano Families in the Urban Southwest, 1848 to the Present.* Notre Dame, Ind.: University of Notre Dame Press, 1984.

Guerin-Gonzalez, Camille. *Mexican Workers and American Dreams: Immigration, Repatriation, and California Farm Labor, 1900–1939.* New Brunswick, N.J.: Rutgers University Press, 1994.

Gutiérrez, David G. *Walls and Mirrors: Mexican Americans, Mexican Immigrants, and the Politics of Ethnicity in the Southwest, 1910–1986.* Berkeley: University of California Press, 1995.

———, ed. *Between Two Worlds: Mexican Immigrants in the United States.* Wilmington, Del.: Scholarly Resources, 1996.

Gutiérrez, Ramón A. *When Jesus Came, the Corn Mothers Went Away: Marriage, Sexuality, and Power in New Mexico, 1500–1846*. Stanford: Stanford University Press, 1991.

Haas, Lisbeth. *Conquests and Historical Identities in California, 1769–1936*. Berkeley: University of California Press, 1995.

Habermas, Jürgen. *Moral Consciousness and Communicative Action*. Translated by Christian Lenhardt and Sherry Weber Nicholsen. Cambridge: MIT Press, 1990.

Hall, Linda B., and Don M. Coerver. *Revolution on the Border: The United States and Mexico, 1910–1920*. Albuquerque: University of New Mexico Press, 1988.

Heinze, Andrew. *Adapting to Abundance*. New York: Columbia University Press, 1990.

Hoffman, Abraham. *Unwanted Mexican-Americans in the Great Depression: Repatriation Pressures, 1929–1939*. Tucson: University of Arizona Press, 1974.

Hondagneu-Sotelo, Pierrette. *Gendered Transitions: Mexican Experience of Im migration*. Berkeley: University of California Press, 1994.

Hurtado, Albert L. *Indian Survival on the California Frontier*. New Haven: Yale University Press, 1988.

Jaramillo, Cleofas M. *Sombras del pasado: Shadows of the Past*. Santa Fe: Ancient City Press, 1941.

Johnson, David R., John A. Booth, and Richard R. Harris. *The Politics of San Antonio: Community, Progress, and Power*. Lincoln: University of Nebraska Press, 1983.

Kanellos, Nicolás. *A History of Hispanic Theatre in the United States: Origins to 1940*. Austin: University of Texas Press, 1990.

Kingsolver, Barbara. *Holding the Line: Women in the Great Arizona Mining Strike of 1983*. Ithaca, N.Y.: ILR Press, 1989.

Kushner, Sam. *Long Road to Delano*. New York: International Publishers, 1975.

Lamphere, Louise, Patricia Zavella, Felipe Gonzales, and Peter B. Evans. *Sunbelt Working Mothers: Reconciling Family and Factory*. Ithaca, N.Y.: Cornell University Press, 1993.

Leonard, Karen Isaksen. *Making Ethnic Choices: California's Punjabi Mexican Americans*. Philadelphia: Temple University Press, 1992.

Levy, Jacques. *César Chávez: Autobiography of La Causa*. New York: W.W. Norton, 1975.

Linden-Ward, Blanche, and Carol Hurd Green. *American Women in the 1960s: Changing the Future*. New York: Twayne, 1993.

Lipsitz, George. *Time Passages: Collective Memory and American Popular Culture*. Minneapolis: University of Minnesota Press, 1990.

Lucas, María Elena. *Forged under the Sun/Forjado bajo del sol: The Life of María Elena Lucas*. Edited by Fran Leeper Buss. Ann Arbor: University of Michigan Press, 1993.

Macias, Anna. *Against All Odds: The Feminist Movement in Mexico to 1940*. Westport, Conn.: Greenwood Press, 1982.

McLean, Robert. *That Mexican! As He Is, North and South of the Rio Grande*. New York: Fleming H. Revell, 1928.

McWilliams, Carey. *Factories in the Field: The Story of Migratory Farm Labor in California*. Boston: Little, Brown, 1935; Santa Barbara: Peregrine, 1971.

————. *North from Mexico.* Philadelphia: J.B. Lippincott, 1949; Westport, Conn.: Greenwood Press, 1968.

Majka, Linda C., and Theo J. *Majka. Farm Workers, Agribusiness, and the State.* Philadelphia: Temple University Press, 1982.

Marchand, Roland. *Advertising the American Dream: Making Way for Modernity, 1920–1940.* Berkeley: University of California Press, 1985.

Marín, Marguerite V. *Social Protest in an Urban Barrio: A Study of the Chicano Movement, 1966–1974.* Lanham, Md.: University Press of America, 1991.

Márquez, Benjamin. *LULAC: The Evolution of a Mexican American Political Organization.* Austin: University of Texas Press, 1993.

Martin, Patricia Preciado. *Songs My Mother Sang to Me: An Oral History of Mexican American Women.* Tucson: University of Arizona Press, 1992.

Martínez, Oscar J. *The Chicanos of El Paso: An Assessment of Progress.* El Paso: Texas Western Press, 1980.

Matsumoto, Valerie J. *Farming the Home Place: A Japanese American Community in California, 1919–1982.* Ithaca: Cornell University Press, 1993.

Mazón, Mauricio. *The Zoot Suit Riots: The Psychology of Symbolic Annihilation.* Austin: University of Texas Press, 1984.

Meyer, Michael, and William Sherman. *The Course of Mexican History.* 5th ed. New York: Oxford University Press, 1995.

Meyerowitz, Joanne J. *Women Adrift: Independent Wage Earners in Chicago, 1880–1930.* Chicago: University of Chicago Press, 1988.

Mirandé, Alfredo, and Evangelina Enríquez. *La Chicana: The Mexican-American Woman.* Chicago: University of Chicago Press, 1979.

Mohanty Chandra Talpade, Ann Russo, and Lourdes Torres, eds. *Third World Women and the Politics of Feminism.* Bloomington: Indiana University Press, 1991.

Monroy, Douglas [Guy]. *Thrown Among Strangers: The Making of Mexican Culture in Frontier California.* Berkeley: University of California Press, 1990.

Mora, Magdalena, and Adelaida R. del Castillo, eds. *Mexican Women in the United States: Struggles Past and Present.* Los Angeles: UCLA Chicano Studies Research Center Publications, 1980.

Moraga, Cherríe. *Loving in the War Years.* Boston: South End Press, 1983.

Moraga, Cherríe, and Gloria Anzaldúa, eds. *This Bridge Called My Back: Writings by Radical Women of Color,* 2nd ed. New York: Kitchen Table Women of Color Press, 1983.

Muñoz Jr., Carlos. *Youth, Identity, and Power: The Chicano Movement.* New York: Verso, 1989.

Orozco, E. C. *Republican Protestantism in Aztlán.* Santa Barbara, Calif.: Petereins Press, 1980.

Padilla, Genaro. *My History Not Yours: The Formation of Mexican American Autobiography.* Madison: University of Wisconsin Press, 1993.

Pascoe, Peggy. *Relations of Rescue: The Search for Female Moral Authority in the American West, 1874–1939.* New York: Oxford University Press, 1990.

Paz, Octavio. *Labyrinth of Solitude: Life and Thought in Mexico.* New York: Grove Press, 1961.

Peiss, Kathy. *Cheap Amusements: Working Women and Leisure in Turn-of-the-Century New York.* Philadelphia: Temple University Press, 1986.

Pesotta, Rose. *Bread Upon the Waters.* New York: Dodd, Mead, 1944.

Ponce, Mary Helen. *Hoyt Street: An Autobiography.* Albuquerque: University of New Mexico Press, 1993.

Pulido, Laura. *Environmentalism and Economic Justice: Two Chicano Struggles in the Southwest.* Tucson: University of Arizona Press, 1996.

Ramos, Juanita, ed. *Compañeras: Latina Lesbians. An Anthology.* New York: Latina Lesbian History Project, 1987.

Rebolledo, Tey Diana. *Women Singing in the Snow: A Cultural Analysis of Chicana Literature.* Tucson: University of Arizona Press, 1995.

Rebolledo, Tey Diana, and Eliana S. Rivero, eds. *Infinite Divisions: An Anthology of Chicana Literature.* Tucson: University of Arizona Press, 1993.

Reed, John. *Insurgent Mexico.* New York: D. Appleton, 1914; New York: International Publishers, 1969.

Ríos-Bustamante, Antonio, ed. *Mexican Immigrant Workers in the U.S.* Los Angeles: UCLA: Chicano Studies Research Center Publications, 1981.

———, and Pedro Castillo. *An Illustrated History of Mexican Los Angeles, 1781–1985.* Los Angeles: UCLA Chicano Studies Research Center Publications, 1986.

Robinson, Cecil. *With Ears of Strangers.* Tucson: University of Arizona Press, 1963.

Rodríguez, Gregorita. *Singing for My Echo.* Santa Fe: Cota Editions, 1987.

Romo, Ricardo. *East Los Angeles: History of a Barrio.* Austin: University of Texas Press, 1983.

Ruiz, Vicki L. *Cannery Women, Cannery Lives: Mexican Women, Unionization, and the California Food Processing Industry, 1930–1950.* Albuquerque: University of New Mexico Press, 1987.

Salas, Elizabeth. *Soldaderas in the Mexican Military: Myth and History.* Austin: University of Texas Press, 1990.

Sánchez, George J. *Becoming Mexican American: Ethnicity, Culture, and Identity in Chicano Los Angeles, 1900–1945.* New York: Oxford University Press, 1993.

Seidman, Steven, ed. *Jürgen Habermas on Society and Politics: A Reader.* Boston: Beacon Press, 1989.

Sheridan, Thomas. *Los Tucsonenses: The Mexican Community in Tucson, 1854–1941.* Tucson: University of Arizona Press, 1986.

Skerry, Peter. *Mexican Americans: The Ambivalent Minority.* New York: The Free Press, 1993.

Soja, Edward. *Postmodern Geographies: The Reassertion of Space in Critical Social Theory.* New York: Verso, 1989.

Soto, Shirlene. *Emergence of the Modern Mexican Woman: Her Participation in Revolution and Struggle for Equality, 1910–1940.* Denver: Arden Press, 1990.

Stein, Walter J. *California and the Dust Bowl Migration.* Westport, Conn.: Greenwood Press, 1973.

Street, Richard Steven. *Organizing for Our Lives: New Voices from Rural Communities.* Portland, Ore.: Newsage Press and California Rural Legal Assistance, 1992.

Tafolla, Carmen. *To Split a Human.* San Antonio, Tex.: Mexican American Cultural Center, 1985.

Tatum, Noreen Dunn. *A Crown of Service.* Nashville, Tenn.: Parthenon Press, 1960.

Taylor, Paul S. *A Mexican-American Frontier: Nueces County, Texas.* Chapel Hill: University of North Carolina Press, 1934.

————. *Mexican Labor in the United States.* 2 vols. Berkeley: University of California Press, 1930–1932; New York: Arno Press, 1969–1970.

Thompson, E. P. *The Making of the English Working Class.* New York: Vintage, 1963.

Tilly, Louise A., and Joan W. Scott. *Women, Work, and Family.* New York: Holt, Rinehart, and Winston, 1978.

Trujillo, Carla, ed. *Chicana Lesbians: The Girls Our Mothers Warned Us About.* Berkeley, Calif.: Third Woman Press, 1991.

Tuck, Ruth. *Not with the Fist: Mexican Americans in a Southwest City.* New York: Harcourt, Brace, 1946; New York: Arno Press, 1974.

Tuttle Jr., William M. *"Daddy's Gone to War": The Second World War in the Lives of American Children.* New York: Oxford University Press, 1993.

Vargas, Zaragosa. *Proletarians of the North: A History of Mexican Industrial Workers in Detroit and the Midwest, 1917–1933.* Berkeley: University of California Press, 1993.

Weber, Devra. *Dark Sweat, White Gold: California Farm Workers, Cotton, and the New Deal.* Berkeley: University of California Press, 1994.

Welty, Eudora. *One Writer's Beginnings.* Cambridge: Harvard University Press, 1983.

Whaley, Charlotte. *Nina Otero-Warren of Santa Fe.* Albuquerque: University of New Mexico Press, 1994.

White, Richard. *"It's Your Misfortune and None of My Own": A New History of the American West.* Norman: University of Oklahoma Press, 1991.

Williams, Norma. *The Mexican-American Family: Tradition and Change.* Boston: G.K. Hall, 1990.

Wilson, Michael. *Salt of the Earth.* Screenplay, with commentary by Deborah Silverton Rosenfelt. Old Westbury, N.Y.: Feminist Press, 1978.

Yung, Judy. *Unbound Feet: A Social History of Chinese Women in San Francisco.* Berkeley: University of California Press, 1995.

Zamora, Emilio. *The World of the Mexican Worker in Texas.* College Station, Tex.: Texas A&M Press, 1993.

Zavella, Patricia. *Women's Work and Chicano Families: Cannery Workers of the Santa Clara Valley.* Ithaca: Cornell University Press, 1987.

Yohn, Susan. *A Contest of Faiths: Missionary Women and Pluralism in the American Southwest.* Ithaca: Cornell University Press, 1995.

Articles

Alexander, Ruth. "'The Only Thing I Wanted Was Freedom': Wayward Girls in New York, 1900–1930." In *Small Worlds: Children and Adolescents in America,* edited by Elliot West and Paula Petrik. Lawrence: University of Kansas Press, 1992.

Allen, Ruth. "Mexican Peon Women in Texas." *Sociology and Social Research* 16 (1931–1932): 131–142.

"Americanization Notes." *Arizona Teacher and Home Journal* 11 (January 1993): 26.

Arrizón, Alicia. "Monica Palacios: 'Latin Lezbo Comic.'" *Crossroads,* no. 31 (May 1993): 25.

Baer, Barbara, and Glenna Matthews. "Women of the Boycott." In *America's*

Working Women: A Documentary History, 1600 to the Present, edited by Rosalyn Baxandall and Susan Reverby, 363–72. New York: Vintage, 1976.

Betten, Neil, and Raymond A. Mohl. "From Discrimination to Repatriation: Mexican Life in Gary, Indiana, during the Great Depression." In *The Chicano,* edited by Norris Hundley, 124–42. Santa Barbara: ABC-Clio Press, 1975.

Booth, John A. "Political Change in San Antonio, 1970–82: Toward Decay or Democracy." In *The Politics of San Antonio: Community, Progress, and Power,* edited by David R. Johnson, John A. Booth, and Richard R. Harris, 193–211. Lincoln: University of Nebraska Press, 1983.

"A Boycott to Aid Garment Workers." *Business Week,* 26 August 1972: 53–54.

Brooks, James F. "'This evil extends especially to the feminine sex': Captivity and Identity in New Mexico, 1700–1847." In *Writing the Range: Race, Class, and Culture in the Women's West,* edited by Susan Armitage and Elizabeth Jameson. Norman: University of Oklahoma Press, 1997, 97–121.

Calderón, Roberto R. "Union, Paz y Trabajo: Laredo's Mutual Aid Societies, 1890s." In *Historical Scholarship in Texas: Proceedings of the Mexicans in Texas History Conference,* edited by Emilio Zamora, Cynthia E. Orozco, and Rudolfo Rocha. Austin: University of Texas Press, forthcoming.

Calderón, Roberto R., and Emilio Zamora. "Manuela Solis Sager and Emma Tenayuca: A Tribute." In *Chicana Voices: Intersections of Class, Race, and Gender,* edited by Teresa Córdova, Norma Cantú, Gilberto Cardenas, Juan García, and Christine M. Sierra, 30–41. Austin: University of Texas Center for Mexican American Studies, 1986.

Candelaria, Cordelia. "La Malinche, Feminist Prototype." *Frontiers* 5:2 (Summer 1980): 1–6.

Cantú, Norma. "Women, Then and Now: An Analysis of the Adelita Image Versus the Chicana as Political Writer and Philosopher." In *Chicana Voices,* eds. Córdova et al., 8–10.

Carroll, Raymond G. "The Alien on Relief." *Saturday Evening Post,* 11 January 1936, 16–17, 100–103.

Castañeda, Antonia. "Sexual Violence in the Politics and Policies of Conquest: Amerindian Women and the Spanish Conquest of Alta California." In *Building with Our Hands: New Directions in Chicana Studies,* edited by Adela de la Torre and Beatríz M. Pesquera, 15–33. Berkeley: University of California Press, 1993.

Castañeda, Irene. "Personal Chronicle of Crystal City: Part II." In *Literatura Chicana,* edited by Antonia Castañeda, Tomás Ybarra-Frausto, and Joseph Sommers, 243–49. Englewood Cliffs, N.J.: Prentice-Hall, 1972.

Cea, Helen Lara. "Notes on the Use of Parish Registers in the Reconstruction of Chicana History in California Prior to 1850." In *Between Borders: Essays on Mexicana/Chicana History,* edited by Adelaida R. del Castillo, 131–60. Encino, Calif.: Floricanto Press, 1990.

Chabram-Dernesesian, Angie. "I Throw Punches for My Race, But I Don't Want to Be a Man: Writing Us Chica-nos (Girl, Us)/Chicanas—Into the Movement Script." In *Cultural Studies,* edited by Lawrence Grossberg, Cary Nelson, and Paula A. Treichler, 81–95. New York: Routledge, 1992.

Chacón, Ramón. "The 1933 San Joaquin Valley Cotton Strike: Strikebreaking Activities in California Agriculture." In *Work, Family, Sex Roles, Language,* edited by Mario Barrera, Albert Camarillo, and Francisco Hernández, 33–70. Berkeley, Calif.: Tonatiuth-Quinto Sol, 1980.

Chávez, Carmen R. "Coming of Age During the War: Reminiscences of an Albuquerque Hispana." *New Mexico Historical Review* 70, no. 4 (October 1995): 383–97.

Chávez, Jennie V. "Women of the Mexican American Movement." *Mademoiselle,* April 1972: 82, 150–52.

Chrystos. "I Don't Understand Those Who Have Turned Away From Me." In *This Bridge Called My Back: Writings by Radical Women of Color,* 2nd ed., edited by Cherríe Moraga and Gloria Anzaldúa, 68–70. New York: Kitchen Table Women of Color Press, 1983.

Córdova, Teresa. "Roots and Resistance: The Emergent Writings of Twenty Years of Chicana Feminist Struggle," in *Handbook of Hispanic Cultures in the United States: Sociology,* ed. Félix Padilla, 175–202. Houston: Arte Publico Press, 1994.

Coyle, Laurie, Gail Hershatter, and Emily Honig. "Women at Farah: An Unfinished Story." In *Mexican Women in the United States: Struggles Past and Present,* edited by Magdalena Mora and Adelaida R. del Castillo, 117–43. Los Angeles: UCLA Chicano Studies Research Center Publications, 1980.

Crocker, Ruth Hutchinson. "Gary Mexicans and 'Christian Americanization': A Study in Cultural Conflict." In *Forging Community: The Latino Experience in Northwest Indiana, 1919–1975,* edited by James B. Lane and Edward J. Escobar, 115–34. Chicago: Cattails Press, 1987.

Culhane, John. "Hero Street, U.S.A." *Reader's Digest,* May 1985: 81–86.

Del Castillo, Adelaida R. "Malintzin Tenépal: A Preliminary Look into a New Perspective." In *Essays on La Mujer,* edited by Rosaura Sánchez and Rosa Martínez Cruz, 124–49. Los Angeles: UCLA: Chicano Studies Center Publications, 1977.

Durón, Clementina. "Mexican Women and Labor Conflict in Los Angeles: The ILGWU Dressmakers Strike of 1933." *Aztlán* 15:1 (1984): 145–61.

Escobar, Edward J. "The Dialectics of Repression: The Los Angeles Police Department and the Chicano Movement, 1968–1971." *Journal of American History* 79, no. 4 (March 1993): 1483–1514.

Ferree, Myra Marx. "Working Class Feminism: A Consideration of the Consequences of Employment." *Sociological Quarterly* 21 (Spring 1980): 173–84.

Friaz, Guadalupe M. "'I Want to be Treated as an Equal': Testimony from a Latina Union Activist." *Aztlán* 20: 1–2 (1991): 195–202.

Gamboa, Erasmo. "Oregon's Hispanic Heritage." *Oregon Humanities,* Summer 1992: 2–7.

García, Alma M. "The Development of Chicana Feminist Discourse, 1970–1980." In *Unequal Sisters: A Multicultural Reader in U.S. Women's History,* edited by Vicki L. Ruiz and Ellen Carol DuBois, 531–44. 2nd ed. New York: Routledge, 1994.

García, John. "Ethnicity and Chicanos: Ethnic Identification, Identity, and Consciousness." *Hispanic Journal of Behavioral Sciences* 4 (1982): 295–314.

Garis, Roy L. "The Mexican Invasion." *Saturday Evening Post,* 19 April 1930: 43–44.

Gaspar de Alba, Alicia R. "Literary Wetback." In *Infinite Divisions: An Anthology of Chicana Literature,* edited by Tey Diana Rebolledo and Eliana S. Rivero, 288–92. Tucson: University of Arizona Press, 1993.

———. "Tortillerismo: Work by Chicana Lesbians." *Signs: Journal of Women in Culture and Society* 18, no 4 (Summer 1993): 956–63.

Glenn, Evelyn Nakano. "From Servitude to Service Work: Historical Continuities in the Racial Division of Paid Reproductive Labor." In *Unequal Sisters,* eds. Ruiz and DuBois, 405–35. 2nd ed.

Goldsmith, Raquel Rubio. "Shipwrecked in the Desert: A Short History of the Mexican Sister of the House of the Providence in Douglas, Arizona, 1927–1949." In *Women on the U.S.-Mexican Border: Responses to Change,* edited by Vicki L. Ruiz and Susan Tiano, 177–95. Boston: Allen and Unwin, 1987.

Gonzales, Philip B. "Spanish Heritage and Ethnic Protest in New Mexico: The Anti-Fraternity Bill of 1933." *New Mexico Historical Review* 61 (October 1986): 281–98.

González, Deena J. "La Tules of Image and Reality: Euro-American Attitudes and Legend Formation on a Spanish-Mexican Frontier." In *Building with Our Hands,* eds. de la Torre and Pesquera, 75–90.

González, María C. "Cultural Conflict: Introducing the Queer in Mexican-American Literature Classes." In *Tilting the Tower: Teaching/Lesbians in the Queer Nineties,* edited by Linda Garber, 56–62. New York: Routledge, 1994.

González, Rosalinda M. "Chicanas and Mexican Immigrant Families, 1920–1940: Women's Subordination and Family Exploitation." In *Decades of Discontent: The Women's Movement, 1920–1940,* edited by Lois Scharf and Joan M. Jensen, 59–83. Westport, Conn.: Greenwood Press, 1983.

González, Sylvia. "The White Feminist Movement: The Chicana Perspective." *Social Science Journal* 14, no. 2 (April 1977): 67–76.

Green, Rayna. "The Pocahontas Perplex." In *Unequal Sisters: A Multicultural Reader in U.S. Women's History,* edited by Ellen Carol DuBois and Vicki L. Ruiz, 15–21. New York: Routledge, 1990.

Gutiérrez, Ramón A. "Community, Patriarchy, and Individualism: The Politics of Chicano History and the Dream of Equality." *American Quarterly* 45 (March 1993): 44–72.

———. "Honor, Ideology, and Class: Gender Domination in New Mexico, 1690–1846." *Latin American Perspectives* 12 (Winter 1985): 81–104.

Hernández, Antonia. "Chicanas and the Issue of Involuntary Sterilization: Reforms Needed to Protect Informed Consent." *Chicano Law Review* 3:3 (1976): 3–37.

Hernández, Inés. "Sara Estela Ramírez: Sembradora." *Legacy* 6:1 (1989): 13–26.

Jensen, Joan M. "'Disenfranchisement Is a Disgrace': Women and Politics in New Mexico, 1900–1940." In *New Mexico Women: Intercultural Perspectives,* edited by Joan M. Jensen and Darlis A. Miller, 301–31. Albuquerque: University of New Mexico Press, 1986.

———. "Pioneers in Politics." *El Palacio* 92 (Summer-Fall 1986): 14–19.

Lecompte, Janet. "The Independent Women of Hispanic New Mexico, 1821–1846." In *New Mexico Women,* eds. Jensen and Miller, 71–93.

Ledesma, Irene. "Texas Newspapers and Chicana Workers' Activism, 1919–1974." *Western Historical Quarterly* 26, no. 3 (Autumn 1995): 309–31.

Limón, José E. "El Primer Congreso Mexicanista de 1911: A Precursor to Contemporary Chicanismo." *Aztlán* 5, nos. 1–2 (Spring and Fall 1974): 85–117.

Longauex y Vasquez, Enriqueta. "The Mexican-American Woman." In *Sisterhood Is Powerful: An Anthology of Writings from the Women's Liberation Movement,* edited by Robin Morgan, 379–84. New York: Vintage, 1970.

Longmore, Wilson T., and Homer L. Hitt. "A Demographic Analysis of First and

Second Generation Mexican Population of the United States: 1930." *Southwestern Social Science Quarterly* 24 (1943): 138–48.

López, Sonia A. "The Role of the Chicana Within the Student Movement. In *Essays on La Mujer,* eds. Sánchez and Martínez Cruz, 16–29.

López, Ronald W. "The El Monte Berry Strike of 1933." *Aztlán* 1 (Spring 1970): 101–14.

Lorimer, George Horace. "The Mexican Conquest." *Saturday Evening Post,* (June 22, 1929): 20.

Malley, Deborah DeWitt. "How the Union Beat Willie Farah." *Fortune,* (August 1974): 164–67, 238, 240.

Martínez, Elizabeth. "Chingón Politics Die Hard: Reflections on the First Chicano Activists Reunion." *Z Magazine,* (April 1990): 46–50.

———. "Levi's, Button Up your Fly: Your Greed is Showing." *Z Magazine,* (January 1993): 22–27.

Martínez, Oscar J. "On the Size of the Chicano Population: New Estimates, 1850–1900." *Aztlán,* 6 (1975): 43–67.

Matsumoto, Valerie J. "Desperately Seeking 'Deidre': Gender Roles, Multicultural Relations, and Nisei Women Writers of the 1930s." *Frontiers: A Journal of Women's Studies* 12 (1991): 19–32.

Melville, Margarita B. "Selective Acculturation of Female Mexican Migrants." In *Twice a Minority: Mexican-American Women,* edited by Margarita B. Melville, 155–63. St. Louis: C.V. Mosby, 1980.

Meyer, Connie. "Concha Ortiz y Pino de Kleven." *Century: A Southwest Journal of Observation and Opinion* (July, 15, 1981): 14–17.

Monroy, Douglas [Guy]. "La Costura en Los Angeles, 1933–1939: The ILGWU and the Politics of Domination." In *Mexican Women in the United States,* eds. Mora and del Castillo, 171–78.

———. "An Essay on Understanding the Work Experiences of Mexicans in Southern California, 1900–1939." *Aztlán* 12 (March 1981): 59–74.

Mora, Magdalena. "The Role of Mexican Women in the Richmond Tolteca Strike." In *Mexican Immigrant Workers in the U.S.,* edited by Antonio Ríos-Bustamante, 111–17. Los Angeles: UCLA Chicano Studies Research Center Publications, 1981.

Moraga, Cherríe. "La Güera." In *Bridge,* 2nd ed., eds. Moraga and Anzaldúa, 27–34.

Moreno, Luisa. "Caravans of Sorrow: Noncitizen Americans of the Southwest." In *Between Two Worlds: Mexican Immigrants in the United States,* edited by David G. Gutierrez, 119–23. Wilmington, Del.: Scholarly Resources, 1996.

Murphy, Mary. "Bootlegging Mothers and Drinking Daughters: Gender and Prohibition in Butte, Montana." *American Quarterly* 46:2 (June 1994): 174–94.

Nelson-Cisneros, Victor B. "UCAPAWA and Chicanos in California: The Farm Worker Period, 1937–1940." *Aztlán* 7 (Fall 1976): 453–77.

Orozco, Cynthia E. "Alicia Dickerson Montemayor: The Feminist Challenge to the League of United Latin American Citizens, Family Ideology, and Mexican American Politics in Texas in the 1930s." In *Writing the Range,* eds. Armitage and Jameson. Norman: University of Oklahoma Press, 1997, 435–56.

———. "Beyond Machismo, La Familia, and Ladies Auxiliaries: A Historiography of Mexican-Origin Women's Participation in Voluntary Associations and Politics in the United States, 1870–1990." *Renato Rosaldo Lecture Series Monograph* 10 (1992–93): 37–77.

————. "Hernández, María L. de." In *Handbook of Texas*, 572–73. Austin: Texas State Historical Association, 1996.

————. "Mujeres Por La Raza." In *Handbook of Texas*, 878–79.

————. "Sexism in Chicano Studies and the Community." In *Chicana Voices*, eds. Teresa Córdova et al., 11–18.

Pérez, Emma M. "A la Mujer: A Critique of the Partido Liberal Mexicano's Gender Ideology on Women." In *Between Borders*, ed. del Castillo, 459–82.

————. "Sexuality and Discourse: Notes from a Chicana Survivor." In *Chicana Lesbians: The Girls Our Mothers Warned Us About*, edited by Carla Trujillo, 159–84. Berkeley: Third Woman Press, 1991.

Pesquera, Beatríz M. "'In the Beginning He Wouldn't Lift Even a Spoon': The Division of Household Labor." In *Building with Our Hands*, eds. de la Torre and Pesquera, 181–95.

————. "Work Gave Me a Lot of Confianza: Chicanas' Work Commitment and Work Identity." *Aztlán* 20:1–2 (1991): 97–116.

Plotkin, Sidney. "Democratic Change in the Urban Political Economy: San Antonio's Edwards Aquifer Controversy." In *The Politics of San Antonio*, eds. Johnson, et al., 157–74.

Pusey, Allen. "Clothes Made the Man." *Texas Monthly*, (June 1977): 133–37.

Reynolds, Jean. "Mona Benítez Piña: An Oral History of a Chicana Farmworker in Phoenix." *Palo Verde* 3, no. 1 (Spring 1995): 65–75.

Roberts, Kenneth L. "The Docile Mexican." *Saturday Evening Post*, (March 10, 1928): 39, 41, 165–66.

————. "Wet and Other Mexicans." *Saturday Evening Post* (February 4, 1928): 10–11, 137–38, 141–42, 146.

Romero, Mary, and Eric Margolis. "Tending the Beets: Campesinas and the Great Western Sugar Company." *Revista Mujeres* (June 2, 1985): 17–27.

Romo, Ricardo. "Mexican Americans in the New West." In *The Twentieth Century West*, edited by Gerald D. Nash and Richard Etulain, 123–45. Albuquerque: University of New Mexico Press, 1989.

Rosales, F. Arturo. "Shifting Self-Perception and Ethnic Consciousness among Mexicans in Houston, 1908–1946." *Aztlán* 16 (1985): 71–94.

Rose, Margaret [Eleanor]. "The Community Service Organization, 1947–1962." In *Not June Cleaver: Women and Gender in Postwar America, 1945–1960*, edited by Joanne Meyerowitz, 177–200. Philadelphia: Temple University Press, 1994.

Ruiz, Vicki L. "'And Miles to Go . . .': Mexican Women and Work, 1930–1985." In *Western Women: Their Land, Their Lives*, edited by Lillian Schlissel, Vicki L. Ruiz, and Janice Monk, 117–36. Albuquerque: University of New Mexico Press, 1988.

————. "By the Day or the Week: Mexicana Domestic Workers in El Paso." In *Women on the U.S.–Mexican Border: Responses to Change*, edited by Vicki L. Ruiz and Susan Tiano, 61–76. Boston: Allen and Unwin, 1987.

————. "Dead Ends or Gold Mines? Using Missionary Records in Mexican American Women's History." *Frontiers: A Journal of Women's Studies*, 12, no. 1 (1991): 33–56.

————. Foreword to *Songs My Mother Sang to Me: An Oral History of Mexican American Women*, by Patricia Preciado Martin. Tucson: University of Arizona Press, 1992.

————. "Moreno, Luisa." In *Encyclopedia of the American West*, edited by Charles Philips and Alan Axelrod, 1030. New York: Macmillan, 1996.

———— "Oral History and La Mujer: The Rosa Guerrero Story." In *Women on the U.S.-Mexico Border,* eds. Ruiz and Tiano, 219–31.

————. "A Promise Fulfilled: Mexicana Cannery Workers of Southern California." *Pacific Historian* 30 (Summer 1986): 50–61.

————. "'Star Struck': Acculturation, Adolescence, and the Mexican American Woman, 1920–1940." In *Small Worlds,* eds. West and Petrik, 61–80.

————. "Teaching Chicano/American History: Goals and Methods." *History Teacher* 20 (February 1987): 67–77.

Salas, Elizabeth. "Ethnicity, Gender and Divorce: Issues in the 1922 Campaign by Adelina Otero-Warren for the U.S. House of Representatives." *New Mexico Historical Review* 70, no. 4 (October 1995): 367–82.

Saldívar-Hull, Sonia. "Feminism on the Border: From Gender Politics to Geopolitics." In *Criticism in the Borderlands: Studies in Chicano Literature, Culture, and Ideology,* edited by Hector Calderón and José David Saldívar, 203–20. Durham: Duke University Press, 1991.

Salmon, Marylynn. "Equality or Submission? Feme Covert Status in Early Pennsylvania." In *Women of America: A History,* edited by Carol Ruth Berkin and Mary Beth Norton, 92–111. Boston: Houghton Mifflin, 1979.

Sánchez, George I., ed. "First Regional Conference on the Education of the Spanish-Speaking People in the Southwest." December, 1945. In *Aspects of the Mexican-American Experience,* with a foreword by Carlos E. Cortés. New York: Arno Press, 1976.

Sánchez, George J. "Go After the Women: Americanization and the Mexican Immigrant Woman, 1915–1929." In *Unequal Sisters,* eds. Ruiz and DuBois, 284–97.

Sandoval, Chela. "U.S. Third World Feminism: The Theory and Method of Oppositional Consciousness in the Postmodern World." *Genders* 4 (1991): 2–24.

San Miguel Jr., Guadalupe. "Culture and Education in the American Southwest: Towards an Explanation of Chicano School Attendance." *Journal of American Ethnic History* 7 (March 1988): 5–21.

Schuyler, Nina. "LA Moms Fight Back." *Progressive* 56: 8 (August 1992): 13.

Segura, Denise A., and Beatríz M. Pesquera. "Beyond Indifference and Antipathy: The Chicana Movement and Chicana Feminist Discourse." *Aztlán* 10:2 (Fall 1988–90): 69–88.

Sekul, Joseph. "Communities Organized for Public Service: Citizen Power and Public Policy in San Antonio." In *The Politics of San Antonio,* eds. Johnson, et al., 175–90.

Soldatenko, María Angelina. "Organizing Latina Garment Workers in Los Angeles." *Aztlán* 20:1-2 (1991): 73–96.

Sosa Riddell, Adaljiza. "Chicanas and El Movimiento." *Aztlán* 5:1 (1974): 155–65.

"South El Paso's Oasis of Care." *paso del norte* 1 (September 1982): 42–43.

Spaulding, Charles S. "The Mexican Strike at El Monte, California." *Sociology and Social Research* 18 (July 1933–August 1934): 571–80.

Tarango, Yolanda. "The Hispanic Woman and Her Role in the Church." *New Theology Review* 3 (November 1990): 56–61.

Taylor, Paul S. "Mexican Women in Los Angeles Industry in 1928." *Aztlán* 11 (Spring 1980): 99–131.

Thompson, Mark. "Threadbare Justice." *California Lawyer* 10 (May 1990): 28–32.

Treviño, Roberto R. "*Prensa y Patria:* The Spanish-Language Press and the Biculturation of the Tejano Middle Class, 1920–1940." *Western Historical Quarterly* 22 (November 1991): 451–72.

Velez I.[bañez], Carlos G. "The Nonconsenting Sterilization of Mexican Women in Los Angeles: Issues of Psychocultural Rupture and Legal Redress in Paternalistic Behavioral Environments." In *Twice a Minority,* ed. Melville, 235–48.

Veyna, Angelina. "'It is My Last Wish that . . .': A Look at Colonial Nuevo Mexicanas through Their Testaments." In *Building with Our Hands,* eds. de la Torre and Pesquera, 91–108.

Vidal, Mirta. "New Voice of La Raza: Chicanas Speak Out." *International Socialist Review* 32, no. 9 (October 1971): 7–9, 31–33.

"Vignette: Concha Ortiz y Pino de Kleven." *The Santa Fean* (December 1974–1975): 18–19.

Weber, Devra. "*Raiz Fuerte:* Oral History and Mexicana Farmworkers." In *Unequal Sisters,* eds. Ruiz and DuBois, 395–404. 2nd ed.

Weinberg, Sydney Stahl. "The Treatment of Women in Immigration History: A Call for Change." In *Seeking Common Ground: Multidisciplinary Studies of Immigrant Women in the United States,* ed. Deborah Gabaccia, 3–22. Westport, Conn.: Greenwood Press, 1992.

Wollenberg, Charles. "Working on El Traque: The Pacific Electric Strike of 1903." In *The Chicano,* ed. Hundley, 96–107.

Zambrana, Ruth. "A Walk in Two Worlds." *Social Welfare* 1 (March 1986): 101–102.

Zamora Jr., Emilio. "Sara Estela Ramírez: Una Rosa Roja en el Movimiento." In *Mexican Women in the United States,* eds. Mora and del Castillo, 163–69.

Zinn, Maxine Baca. "Political Familialism: Toward Sex Role Equality in Chicano Families." *Aztlán* 8, no. 1 (1975): 13–26.

Dissertations, Theses, and Unpublished Papers

Año Nuevo Kerr, Louise. "The Chicano Experience in Chicago, 1920–1970." Ph.D. diss., University of Illinois, Chicago Circle, 1976.

Arredondo, Gabriela. "'Equality' for All: Americanization of Mexican Immigrant Women in Los Angeles and San Francisco through Newspaper Advertising, 1920–1935." M.A. seminar paper, San Francisco State University, 1991.

Becerra, Victor. "The Untold Story of Chávez Ravine." Unpublished paper, History Research Holdings, Chicano Studies Library, University of California, Los Angeles, California.

Buriel, Ray. "The Cultural Adaptation of Hispanic Immigrants in the Citrus Belt, 1920–Present." Unpublished paper.

Cadena, Gilbert, and Lara Medina. "Liberation Theology and Social Change: Chicanas and Chicanos in the Catholic Church." Unpublished manuscript.

Camarillo, Albert. "Mexican American Urban History in Comparative Ethnic Perspective." Distinguished Speakers series, University of California, Davis, 26 January 1987.

Castañeda, Antonia I. "Presidarias y Pobladoras: Spanish-Mexican Women in Frontier Monterey, Alta California, 1770–1821." Ph.D. diss., Stanford University, 1990.

Cervantes, Angel. "'The Beginning of Our Being . . .': The East Los Angeles Rebellion of 1968." M.A. thesis, Claremont Graduate School, 1996.

Chambers, Clarke Alexander. "A Comparative Study of Farm Organizations in California during the Depression Years, 1929–1941." Ph.D. diss., University of California, Berkeley, 1950.

Chávez, Ernesto. "Creating *Aztlán*: The Chicano Student Movement in Los Angeles, 1966–1978." Ph.D. diss., University of California, Los Angeles, 1994.

Chávez, Marisela R. "Nuevomexicana Cultural Invention as Cultural Resistance: Fabiola Cabeza de Baca, Cleofas M. Jaramillo, and Nina Otero Warren, 1880–1955." Graduate seminar paper, Arizona State University, 1996.

————. "'We Lived and Breathed and Worked the Movement': Women in El Centro Acción Social Autónomo (CASA), 1975–1978." Graduate seminar paper, Arizona State University, 1996.

Chávez, Miroslava. "Don't Kill Me Sister: Gender, Race and Power in the Murder Trial of Guadalupe Trujillo in Mexican Los Angeles." Paper presented at the Berkshire Conference on the History of Women, University of North Carolina, Chapel Hill, 7–9 June 1996.

Dewanto, Nirwan."American Kitsch and Indonesian Culture: A Sketch." Paper presented at symposium, American Studies in the Asia-Pacific Region: The Impact of American Popular Culture on Social Transformation in Asian Countries, Tokyo, Japan, 3 April 1992.

Douglas, Helen. "The Conflict of Cultures in First Generation Mexicans in Santa Ana, California." M.A. thesis, University of Southern California, 1928.

Espino, Virginia. "Women Sterilized As You Give Birth: Forced Sterilization and Chicana Resistance in the 1970s." Paper presented at the National Association for Chicana and Chicano Studies, Chicago, 21–23 March 1996.

Garcílazo, Jeff. "Taylorism, Patriarchy and *Traqueros*: Mexican Railroad Workers in the United States, 1880–1930." Unpublished paper, 1995.

————. *"Traqueros*: Mexican Railroad Workers in the United States, 1871–1930." Ph.D. diss., University of California, Santa Barbara, 1995.

González, Deena J. "Spanish-Mexican Women on the Santa Fe Frontier: Patterns of Their Resistance and Accommodation, 1820–1880." Ph.D. diss., University of California, Berkeley, 1985.

Hymer, Evangeline. *A Study of the Social Attitudes of Adult Mexican Immigrants in Los Angeles and Vicinity, 1923.* San Francisco: R and E Research Associates, 1971.

Jaehn, Thomas. "Concha Ortiz y Pino's Voting Record (field notes based on legislative documents housed at the New Mexico State Records Center and Archive, Santa Fe).

Landolt, Robert Garland. "The Mexican-American Workers of San Antonio, Texas." Ph.D. diss., University of Texas, 1965.

Larson, Anne. "A Comparative Analysis of Proposition 187 and the Immigration Action of 1924." Seminar paper, Claremont Graduate School, Claremont, California, 1994.

López, Yolanda. Panel presentation. Puro Corazón: A Symposium of Chicana Art, Pomona College, Claremont, California, 16 February 1995.

McBane, Margo. "Tale of Two Cities: A Comparative Study of the Citrus Heartlands of Santa Paula and LaVerne." Unpublished paper, 1995.

Malaret, Jesús Francisco. "Of Pachucos, Zoot Suiters, and the Press." Unpublished seminar paper, University of California, Davis, 1992.

Martínez, Rubén. "The Role of the Intellectual Chicano/Latino in Our Communities." Talk given at the Graduate Humanities Center, Claremont Graduate School, Claremont, Calif., 6 October 1993.

Matsumoto, Valerie J. "Growing Up with Jeanette MacDonald and Dear Deidre: Nisei Girls in the 1930s." Unpublished essay, 1991.

Monroy, Douglas Guy. "Mexicanos in Los Angeles, 1930–1941: An Ethnic Group in Relation to Class Forces." Ph.D. diss., University of California, Los Angeles, 1978.

Orozco, Cynthia E. "The Origins of the League of United Latin American Citizens (LULAC) and the Mexican American Civil Rights Movement in Texas with an Analysis of Women's Political Participation in a Gendered Context, 1910–1929." Ph.D. diss., University of California, Los Angeles, 1992.

Pardo, Mary. "Identity and Resistance: Mexican American Women and Grassroots Activism in Two Los Angeles Communities." Ph.D. diss., University of California, Los Angeles, 1990.

Quiñonez, Naomi Helena. "Hijas de la Malinche (Malinche's Daughters): The Development of Social Agency Among Mexican American Women and the Emergence of First Wave Chicana Cultural Production." Ph.D. diss., Claremont Graduate School, 1996.

Rose, Margaret Eleanor. "Women in the United Farm Workers: A Study of Chicana and Mexicana Participation in a Labor Union, 1950 to 1980." Ph.D. diss., University of California, Los Angeles, 1988.

Ruiz, Vicki L. "The Flapper and the Chaperone." Presentation and discussion at the Riverside Municipal Museum, 28 May 1995.

———. "Settlement Houses in El Paso." Audiotape of presentation and discussion at the El Paso Conference on History and the Social Sciences, 24 August 1983, El Paso, Texas.

Segura, Denise A. "Chicanas and Education: Implications for Occupational Mobility," (unpublished paper, 1995).

Sánchez, George J. "The Rise of the Second Generation: The Mexican American Movement." Unpublished paper.

Sánchez-Walker, Marjorie. "Woven Within My Grandmother's Braid: The Biography of a Mexican Immigrant Woman, 1898–1982." M.A. thesis, Washington State University, 1993.

Smith, Clara Gertrude. "The Development of the Mexican People in the Community of Watts." M.A. thesis, University of Southern California, 1933.

Soldatenko, María Angelina. "The Everyday Lives of Latina Garment Workers in Los Angeles: The Convergence of Gender, Race, Class, and Immigration." Ph.D. diss., University of California, Los Angeles, 1992.

Taylor, Quintard. "Exploration and Early Settlement." Unpublished Paper, 1995.

Thurston, Richard G. "Urbanization and Sociocultural Change in a Mexican-American Enclave." Ph.D. diss., University of California, Los Angeles, 1957; reprint, San Francisco: R and E Research Associates, 1974.

Vargas, Zaragosa. "'La Pasionaria de Tejas': Emma Tenayuca, 1930s Mexican American Communist Labor Organizer of San Antonio, Texas." Unpublished paper, 1994.

Villarreal, Mary Ann. "Tejanas of La Raza Unida." Graduate seminar paper, Arizona State University, Tempe, Arizona, 1996.

Newspapers and Newsletters

Advance (newspaper of the Amalgamated Clothing Workers of America)
Agricultural Bulletin (newsletter of the National Council to Aid Agricultural Workers, ca. 1930s)
Albuquerque Journal
Albuquerque Tribune
Chicago Tribune
Chicano Student News
Chicano Studies Newsletter (UC Berkeley)
Chronicle of Philosophy
Daily Worker
El Chicano, (Riverside)
El Paso Herald Post
El Paso Times
El Grito
Encuentro Femenil (Los Angeles)
FTA News (newspaper of Food, Tobacco, Agricultural, and Allied Workers Union).
Guardian (San Francisco)
Hispano America (San Francisco)
La Causa (Los Angeles)
La Crónica
La Gente (UCLA Chicano student paper)
La Opinión (Los Angeles)
La Raza (Los Angeles)
Los Angeles Downtown News
Los Angeles Evening Herald Express
Los Angeles Herald Examiner
Los Angeles Times
Los Angeles Weekly News
Lubbock Avalanche Journal
Magazín
New York Times
Noticiera de M.A.L.C.S. (UC Davis)
People's World
Regeneración (Los Angeles)
San Francisco Examiner
San Francisco News
Santa Fe New Mexican
Sin Fronteras (Los Angeles)
Washington Post

Manuscript Collections

Aguirre Family Papers, Arizona Historical Society Library, Tucson, Arizona.
Arizona Collection, Hayden Library, Arizona State University, Tempe, Arizona.
Chicano Blowouts Collection, Chicano Resource Center, East Los Angeles Public Library, Los Angeles, California.

Chicano Research Collection, Hayden Library, Arizona State University, Tempe, Arizona.

Colección Tloque Nahuaque, Chicano Studies Library, University of California, Santa Barbara, California.

Del Rancho al Barrio Collection, Arizona Historical Society Library, Tucson, Arizona.

Manuel Gamio Collection, Bancroft Library, University of California, Berkeley, California.

La Gente Exhibit, Colorado Historical Society Library, Denver Colorado.

Houchen Files, Houchen Community Center, El Paso, Texas.

George Lambert Collection, Labor Archives, University of Texas, Arlington, Texas.

Russell Lee Collection, Eugene C. Barker History Center, University of Texas, Austin, Texas.

Simon J. Lubin Collection, Bancroft Library, University of California, Berkeley, California.

Carey McWilliams Collection, Special Collections, University of California, Los Angeles, California.

Mexican American Women As Community Builders Photo Exhibit, University Library, Our Lady of Lake University, San Antonio, Texas.

Concha Ortiz y Pino de Kleven Collection, Special Collections, General Libraries, University of New Mexico.

Esther Pérez Papers, Cassiano-Pérez Collection, Daughters of the Republic of Texas Library at the Alamo, San Antonio, Texas.

Rio Grande Historical Collection, University Library, New Mexico State University, Las Cruces, New Mexico.

Carmen Tafolla Papers, Benson Library, University of Texas, Austin, Texas.

Paul S. Taylor Collection, Bancroft Library, University of California, Berkeley, California.

Women in New Mexico Collection, Special Collections, General Libraries, University of New Mexico.

Correspondence and Miscellaneous Documents

Goldman, Shifra M. to Vicki L. Ruiz, 26 June 1984.

Houghteling, Tom, to Vicki L. Ruiz, 24 December 1990.

Metzger, Sidney, Bishop of San Antonio to the Bishop of Rochester, 31 October 1972. In *Viva la Huelga: Farah Strike Bulletin*, no. 15. McWilliams, Cary to Louis Adamic, dated 3 October 1937, *Adamic File*, Carey McWilliams Collection, Special Collections, University of California, Los Angeles, California.

Moreno, Luisa, to Vicki L. Ruiz, 12 August 1983.

MALCS chapter members and concerned students at UCLA to Professor Ruth Schwarz, Department of Film and Television, UCLA dated 20 June 1989

American Immigration Control. "U.S. Citizen Opinion Poll on America's Illegal Alien Crisis" (1995).

"Blood on the Cotton," UCAPAWA (United Cannery, Agricultural, Packing, and Allied Workers of America) pamphlet (1939).

California State University. Heller Committee for Research in Social Economics and Constantine María Panunzio. *How Mexicans Earn and Live: A Study of the Incomes and Expenditures of One Hundred Mexican Families in San Diego, California.* University of California Publications in Economics, vol. 13: Cost of Living Studies, no. 5. Berkeley: University of California Press, 1933.

Carmejo, Antonio, ed. *Why a Chicano Party? Why Chicano Studies?* Pamphlet. New York: Pathfinder Press, 1970.

Chronicle of Philanthropy, July 24, 1990.

Flyer announcing a march against Proposition 187 scheduled for October 16 in Los Angeles.

National Latino Communications Center. "Chicano! History of the Mexican American Civil Rights Movement." Proposal submitted to the National Endowment of the Humanities, August 1994.

National Women's History Project. *Las Mujeres: Mexican American/Chicana Women.* Curriculum Booklet. Windsor, Calif.: National Women's History Project, 1991.

Poling, Paul Newton. *For the Defense of the Farah Workers.* Pamphlet. El Paso, Tex.: Farah Manufacturing Co., ca. 1973.

Portland (Ore.) Strike Support Committee. "The Farah Strike: Our Struggle, Too." Pamphlet, 1973.

"Raíz Fuerte que no se Arranca," memorial program in honor of Magdalena Mora. June 13, 1981.

San Francisco Bay Area Farah Strike Support Committee. "Chicanos Strike at Farah." Pamphlet, 1974.

Sí Magazine (Summer 1996): 6.

Viva la Huelga: Farah Strike Bulletins New York: Amalgamated Clothing Workers of America.

Steinbeck, John. *Their Blood Is Strong.* Pamphlet. San Francisco: Simon J. Lubin Society, 1938.

United States Chamber of Commerce. Immigration Committee. *Mexican Immigration.* Washington, D.C.: U.S. Chamber of Commerce, 1930.

Winkler, Karen. "Scholars Say Issues of Diversity Have Revolutionized Chicano Studies." *Chronicle of Higher Education,* 26 September 1990

Interviews

Acosta, Lucy. Interview no. 653 by Mario T. García, 28 October 1982. Institute of Oral History, University of Texas, El Paso, Texas.

Arredondo, María. Interview by Carolyn Arredondo, 19 March 1986.

Barcena, Angela. Interview by Oscar Martínez, Mario Galdos, and Virgilio Sánchez, 2 August 1979. Institute of Oral History, University of Texas, El Paso, Texas.

Buriel, Eusebia. Interview by Vicki L. Ruiz, 16 January 1995.

Buriel, Ray. Interview by Vicki L. Ruiz, 21 December 1994.

Chávez, Elsa [pseud.]. Interview by Vicki L. Ruiz, 19 April 1983.

Clifton, Beatrice Morales. In *Rosie the Riveter Revisited: Women and the World War II Work Experience,* edited by Sherna Berger Gluck, Vol. 8 Long Beach: California State University Long Beach Foundation, 1983.

Escobar, Carmen Bernal. Interview by Vicki L. Ruiz, 11 February 1979.

———. Interview by Vicki L. Ruiz, 15 June 1986.

Estrada, Ruby. Interview by María Hernández, 4 August 1981. The Lives of Arizona Women Oral History Project, Special Collections, Hayden Library, Arizona State University, Tempe, Arizona.

Fierro, María. In *Rosie the Riveter Revisited: Women and the World War II Work Experience,* ed. Gluck, Vol. 12.

Fierro de Bright, Josefina. Interview by Albert Camarillo, 7 August 1977.

García, Alma Araiza. Interview by Vicki L. Ruiz, 27 March 1993.

García, Fernando. Interview by Vicki L. Ruiz, 21 September 1983.

González, Martha. Interview by Vicki L. Ruiz, 8 October 1983.

Healey, Dorothy Ray. Interview by Vicki L. Ruiz, 21 January 1979.

Ibarra, Estella. Interview by Jesusita Ponce, 11 November 1982.

Linares, Clemente. Interview by Lydia Linares Peake, 11 June 1992.

Lucero, Lucy. Interview by Vicki L. Ruiz, 8 October 1983.

Luna, Mary. In *Rosie the Riveter Revisited: Women and the World War II Work Experience,* ed. Gluck, Vol. 20.

Mason, Belén Martínez. In *Rosie the Riveter Revisited: Women and the World War II Work Experience,* ed. Gluck, Vol. 23.

Milligan, Adele Hernández. In *Rosie the Riveter Revisited: Women and the World War II Work Experience,* ed. Gluck, Vol. 26

Moreno, Ernest. Interview by Vicki L. Ruiz, 10 January 1983.

Moreno, Graciela Martínez. Interview by Vicki L. Ruiz, 10 January 1983.

Moreno, Luisa. Interview by Albert Camarillo, 12–13 August 1977.

———. Interview by Vicki L. Ruiz, 27 July 1978,

———. Interview by Vicki L. Ruiz, 4–6 September 1979.

———. Interview by Vicki L. Ruiz, 3 August 1984.

Mount, Julia Luna. Interview by Vicki L. Ruiz, 17 November 1983.

Mulligan, Rose Escheverría. In *Rosie the Riveter Revisited: Women and the World War II Work Experience,* ed. Gluck, Vol. 27.

"Elizabeth Nicholas: Working in the California Canneries." [interview conducted by Ann Baxandall Krooth and Jaclyn Greenberg]. *Harvest Quarterly,* nos. 3–4, September–December 1976.

Ortiz y Pino de Kleven, Concha. Interview by Vicki L. Ruiz, 16 September 1988.

R., Lucia. Interview by Margarita C., December 1981.

Romero, María del Carmen. Interview by Vicki L. Ruiz, 8 January 1983.

Ruiz, Erminia. Interview by Vicki L. Ruiz, 30 July 1990.

———. Interview by Vicki L. Ruiz, 18 February 1993.

Shelit, Alicia Mendeola. In *Rosie the Riveter Revisited: Women and the World War II Work Experience,* ed. Gluck, Vol. 37.

"Un cuento—una vida." Unpublished interview with Gregoria Sosa conducted by Adaljiza Sosa Riddell.

Torres, Jesusita. Interview by Vicki L. Ruiz, 8 January 1993.

Ybarra, María. Interview by David Pérez, 1 December 1990.

Poems

Blake, William. "Auguries of Innocence." In *Poems of William Blake,* edited by W. H. Stevenson, 585. London: Longman, 1971.

Carillo, Jo. "And When You Leave, Take Your Pictures with You." In *Bridge*, eds. Moraga and Anzaldúa, 63–64.

Corpi, Lucha. "Marina Mother." Translated by Catherine Rodríguez-Nieto. In *Infinite Divisions*, eds. Rebolledo and Rivero, 196–97.

Correa, Viola. "La Nueva Chicana" In Córdova, "Roots and Resistance," in *Handbook of Hispanic Cultures*, ed. Padilla, 182.

Flores, Isabel. "I Remember." *El Grito* 7 (September 1973): 80.

Frost, Robert. "Stopping by the Woods on a Snowy Evening." In *Complete Poems of Robert Frost*. New York: Holt, Rinehart, and Winston, 1958.

Lucero Trujillo, Marcela Christine. "Machismo Is Part of Our Culture," In Gutiérrez, "Community, Patriarchy, and Individualism," 48.

Mora, Pat. "University Avenue." In *Borders*. Houston: Arte Público Press, 1986.

Quiñonez, Naomi. "The Confession." In *Sueño de Colibrí/ Hummingbird Dream*. Los Angeles: West End Press, 1985.

Ramírez, Sara Estela. "Surge! (A la mujer)." In Inés Hernández, "Sara Estela Ramírez: Sembradora," *Legacy*, 6:1 (1989):22.

Sosa Riddell, Adaljiza. "Como Duele." In *Infinite Divisions*, eds. Rebolledo and Rivero, 213–15.

Tafolla, Carmen. "La Malinche." In *Infinite Divisions*, eds. Rebolledo and Rivero, 198–99.

Visual Media

Adelante Mujeres! Produced by National Women's History Project, 1991. Videocassettes.

A Century of Women. 3 vols. Produced by Turner Home Entertainment, 1994. Videocassettes.

CHICANO! 4 vols. Produced by National Latino Communications Center and Galán Productions, 1996. Videocassettes.

The Lemon Grove Incident. Produced by Paul Espinosa, 1985. Videocassette.

Los Mineros. Produced by Galán Productions, 1991. Videocassette. "Pachuca with a Razor Blade." By Carmen Lomas Garza. Original artwork in the collection of Sonia Saldívar Hull.

Salt of the Earth. Produced by Paul Jarrico, 1954. 16 mm. Available in videocassette.

Index

Abraham Lincoln High School, 103
Acosta, Lucy, 59
Acuña, Rudy, 76, 95, 118, 136
Adamic, Louis, 72
Adelante Mujeres, 139.
African Americans, 78, 103, 141
Aguirre, Julia, 17
Alatorre, Chole, 117. *See also* CASA
Albuquerque Journal, 94. *See also* Ortiz y Pino, Concha
Albuquerque, N. Mex., labor relations in, 137
Alinsky style organizing, 138–39, 141. *See also* Communities Organized for Public Service (COPS)
Allen, Ruth, 14–15
Amalgamated Clothing Workers, 128–29. *See also* Farah textile strike
American Association of University Women, 86
American Committee for the Protection of the Foreign Born (ACPFB), 97. *See also* Moreno, Luisa
American Federation of Labor, 96
American G.I. Forum, 97
American Immigration Control, 147
Americanization, 33, 52. *See also* "Christian Americanization"
Anderson, Karen, 136

Anzaldúa, Gloria, 110, 121–22, 126. *See also This Bridge Called My Back*
Araiza, Eduardo, 31
Armendaríz, Pedro, 111
Arredondo, María, 17, 20
Arrizón, Alicia, xiv–xv
Asociación Nacional México-Americana (ANMA), 102
Associated Farmers, 78–79
Aztec iconography; Chicano nationalism, in, 104; Chicano Student Movement, in, 105–6; El Cerezo Restaurant, 106; "Mi Amor," 109. *See also* La Malinche
Aztecs, the, 87
Aztlán; concept of, 104. *See also* Aztec iconography

Babbitt, C.S., 40–41
Baca, Judy, 123, 150
Bains, Amalia Mesa, 150
Balderrama, Francisco, 11
Bancroft Library, 45
Barcena, Angela, 149–50
Barrera, Mario, 9
Barreras, Catalina, 85. *See also Salt of the Earth*
beauticians, 56
beauty pageants, 55; La Reina de Cinco de Mayo, 56; Orange Queen, 56

Becerra, Victor, 98, 118
Becoming a Writer. See Welty, Eudora
beet harvesting, 11
Belen, N. Mex., 22
Belmont High School, Los Angeles, 104
Bernal, Diana, 84
Betancourt family, 21
Bethelem Steel, Pa., 10
Biberman, Herbert, 85. *See also Salt of the Earth*
Black Student Union (BSU), 103
Blackfoot, Idaho, 10
Blake, William, xvi
Blissfield, Mich., 18
Bonilla, Febe, 47
Bright, John, 96. *See also* Fierro de Bright, Josefina
Brooks, James, xv
Brown v. Board of Education, 84
Buena Park, 4, 30–31
Building With Our Hands, 100
Bullock's [department store], 53
Burciaga, Soledad, 41
Bureau of the Census, 147
Buriel, Eusebia Vásquez de, 10, 22, 24, 66
Buriel, Frederico, 66
Buriel, Ray, 7, 22
Buss, Fran Leeper, 133

Cabeza de Baca, Fabiola, 24. *See also We Fed Them Cactus*
Calderón, Juana, 23
California Rural Legal Assistance, 144
California Sanitary Canning Company (Cal San), 82, 101; labor force at, 80–81; strike at, 81
California Senate UnAmerican Activities Committee, 84
Calpulli, 141–42
Camacho, Irma, 129
Camarillo, Albert, 8–9, 68
Cameron House, San Francisco, 47
Candelaria, Cordelia, 107
Cannery and Agricultural Workers Industrial Union (CAWIU), 74–75
Cannery Women, Cannery Lives, xiii, 15. *See also* Ruiz, Vicki
Cantú, Norma, 111
Cardoso, Lawrence, 9
Carmona, María "Cuca," 144
Carrillo, Jo, 110
Casa Blanca, barrio of, 7
Castañeda, Antonia, xv
Castañeda, Irene, 16–17, 19
Castillo, Ana, 121, 125

Castillo, Aurora, 143
Castro, Rosie, 100
Castro, Sadie, 76
Castro, Sal, 103
Castro, Vicki, 103
Catholic Church; *Calpulli*, 141–42; feminist criticisms of, 108; UFW, support for, 132; warnings against dancing, 65–66; women in, 132–33, 138
Chabram-Dernersesian, Angie, 112
Chamber of Commerce of Los Angeles, 77
Chambers, Pat, 74
Changing Women. See Anderson, Karen
chaperonage. *See* familial protectiveness
Chávez Ravine, 97–98
Chávez, Carmen, 82
Chávez, Cesar, 105, 119, 130. *See also* United Farm Workers
Chávez, Denise, 150
Chávez, Elsa, 33, 38; as Farah textile striker, 127–28, 130–32
Chávez, Ernesto, 116
Chávez, Helen, 134–35
Chávez, Isabel Rodríguez, 118. *See also* CASA
Chávez, Marisela, 118. *See also* CASA
Chávez, Ofilia [sic], 47
Chem Waste, 144
Chicana Action Service Center, 112, 114. *See also* Chicana feminism
Chicana Critical Issues, 123. *See also* MALCS
Chicana feminism; consciousness of, 125; *The Chicana Feminist*, 119; "hierarchy of oppression" in, 122–23; "La Adelita" (*soldadera*), 111; machismo, reaction to, 120; MALCS, 122–23; male reaction to, 109; Mexican-American women and, 99–100; relations between Chicanas, 102, 110; relations between white and Chicana feminists, 110; relations with Chicano nationalists, 111–12; sterilization and Comisíon Femenil Mexicana, 112–14; *This Bridge Called My Back*, 110; Marxism, 100. *See also* Chicana lesbianism
Chicana lesbianism, 100, 121–22
Chicana Lesbians: The Girls Our Mothers Warned Us About, 121–22
Chicana Welfare Rights Organization, 113
Chicana, as feminist, xv; as writer, xv; meaning of term, xiv
Chicanismo, 105

Chicano Moratorium (August 29, 1970), 115

Chicano/a, origin of, 98

Chicano Student Movement, xiv, 97–98, 122; birth of, 102–3; feminism in, 108–9; work for United Farm Workers, 119

Chicano Studies, 105, 112; California State University at Los Angeles program, 103; UCLA, 124

Chiquita, Gloria, 133

"Christian Americanization," 32, 34, 38–39, 41, 46, 50. *See also* Rose Gregory Houchen Settlement House

Cinco de Mayo, 54, 87, 107

Cinderella Ballroom, xvii, 59

Cisneros, Sandra, 150

Ciudad Juárez, Mexico, 3, 31

Clark, Margaret, 68

Coachella Valley, 144

Colton, Calif., 23

Columbine Strike of 1927, 27

Comisíon Femenil Mexicana, 112. *See also* sterilization

commadrazgo; concept of, 4, 16, 102. *See also mutualistas*

Communidades Ecclesiales de Base (CEBs), 141

Communist Party, 74, 84; El Congreso, 95; Texas, 79

Communities Organized for Public Service (COPS), 138–41; Alinksy methods, 138–39; Cisneros, Henry, 139; Community Development Block Grants, 139

Community Redevelopment Agency, 141

Community Service Organization (CSO), 132

Compañeras: Latina Lesbians, 121–22

Congressional Union, 91

consumer ethic, 52, 56–57

contraception, 49, 62

Córdova, Teresa, 125

Corona, Bert, 86, 95, 117

Corpi, Lucha, 107. *See also* La Malinche

Corpus Christi, Tex., 90

Correa, Viola, 100, 156–57

corridos, 62

Cortés, Ernie, Jr., 138, 140

Cortés, Hernán, 106. *See also* La Malinche

Cortez, Beatrice, 140

Cotera, Martha, 115, 119. *See also The Chicana Feminist*

Coyle, Laurie, 128, 130, 132

Cristero Revolt, 8

Cristonomo, Paula, 104

Crocker, Ruth, 44

Crusade for Social Justice, 104

Crystal City, Tex., 115. *See also* La Raza Unida

Cuarón, Mita, 104

cultural coalescence, concept of, xvi, 49–50

curanderas (healers), 25

Curtis, Jesse Judge, 113–14. *See also Madrigal v. Quilligan;* sterilization

Dark Sweat, White Gold, 16. *See also* Weber, Devra

Davis, Mike, 138

de la Torre, Adela. *See Building With Our Hands*

De León, Angel, 96. *See also* Moreno, Luisa, 96

Decker, Caroline, 74

Del Castillo, Adelaida, 106–7, 110, 112

Del Rio, Dolores, 96, 111

Delhi, Calif., 17

Demographics, of Mexican-American women, 147–48; educational levels of, 154; incomes of, 148–49, 155–56; garment workers, 149; occupational distribution of, 153–54

Denver, Colo., 104

Deukmejian, George, 143

Deutsch, Sarah, xv, 27, 44–45. *See also No Separate Refuge*

Dewanto, Nirwan, 52–53

Diaz, Porfiro, 7–8. *See also* Mexican immigration

"dime-a-dance," 14

Dockers jeans. *See* Levi-Strauss

Dodger Stadium, 98

Donald Henderson, 79. *See also* UCAPAWA

Douglas Aircraft, 82–83

Douglas, Ariz., 26

Durango, Mexican state of, 3

Durazo, María Elena, 123, 137–38, 150

Dust Bowl, 30, 77. *See also* "Okie" migration

East Los Angeles high school Chicano student walkout (1968), 103–4

East Valleys Organization (EVO), 141

Eastside Sun, The (East Los Angeles), 83–84

El Buen Pastor, 42, 48. *See also* Rose Gregory Houchen Settlement House

El Centro de Acción Autonoma-Hermandad General de Trabajadores (CASA); Marxism of, 117–18; origins of, 116–17; role of women in, 118–20; *Sin Fronteras,* 117–18
El Chicano, 107–8. *See also* Harrison, Gloria Macias
El Congreso de Pueblos de Hablan Española (Spanish Speaking people's Congress), 73, 79, 96–97; Corona, Bert, role in, 95; empowering women, 97; Fierro de Bright, Josefina, role in, 95, 97; Moreno, Luisa, role in, 94–95; second California convention, 100–101; Quevedo, Eduardo, role in, 95
El Grito, 107. *See also* Martínez, Betita
El Monte Berry Strike of 1933, 75–77
El Monte Hicks Camp Barrio, 76
El Monte Historical Museum, xvi–xvii
El Movimiento Estudantil Chicano de Aztlán, 105, 119
El Nogal Pecan Shellers Union, 79–80
El Paso, Tex., 10, 25, 31–32, 127; Community Chest of, 41; demographics of, 35, U.T. El Paso, 132; *See also* Farah textile strike; Rose Gregory Houchen Settlement House; Segundo Barrio
El Plan de Acción de Calpulli. *See Calpulli*
El Plan de Santa Barbara, 105, 119
El Primer Congreso Mexicanista, 99
El Rebozo, 111
Elko, Nev., 21
Ellis Island, 13
Ellis, Pearl, 33
Empire Zinc, 85. *See also Salt of the Earth*
Encuentro Femenil, 107, 109. *See also* Gómez, Ana Nieto
Escalante, Alicia, 112
Escobar, Carmen Bernal, 59, 61–63, 80, 82. *See also* California Sanitary Canning Company and UCAPAWA
Esparza, Moctezuma, 115
Esparza, Pasquala, 3, 11, 14
Espino, Virginia, 113
Estrada, Ruby, 54, 58–59, 65
Evans, Peter, 137
Evans, Sara, 140
Ewen, Elizabeth, 52, 58

familial oligarchy, 52, 69, 71. *See also* familial protectiveness
familial protectiveness, 58–59; Catholic Church, 66–67; chaperonage, 61, 63–67, 70–71; early marriage as

response to, 60, 61; elopement as response to, 60–61; *vergüenza* (female chastity), 65
family wage economy, 63
Farah textile strike, 127–32; children, role of, 131; Farah Distress Fund, 130; Farah, Willie, 128, 145–46; impact on families, 130; national boycott against, 130–31; picketing controversy at, 128–29; production quotas at, 127; settlement of, 131; *Unidad Para Siempre,* 130, 149; women, role of 131
Farah, Willie. *See* Farah textile strike
Farm Labor Organizing Committee (FLOC), 133
Fernandez, Beatrice, 46–47
Ferreira, Enrique, 88
Fierro de Bright, Josefina, 73, 100; El Congreso, 95, 97; family background, 96; Zoot Suit Riots, witness of, 84. *See also* Bright, John
Fierro, María, 55
Fitzgerald, F. Scott, 55
flappers, 55, xiii
Florence Crittenden Home for Unwed Mothers, 71
Flores, Francisca, 108, 114. *See also* Chicana Action Service Center in East Los Angeles, 114
Flores, Isabel; "I Remember" (poem), 18
Floyd Street Settlement, Dallas, Tex., 47
Food 4 Less, 141
forced repatriation, 29–31
Frances De Pauw School, 42–43
Franco, Minerva, 45
Freeman Clinic, 36, 41
Friendly House, Phoenix, 34
Frost, Robert, 147
Fuentes, Carlos, 106
Fuerza Unida, 149, 151

Gabaccia, Donna, 52
Galisteo, N. Mex., 92
Gallo, Juana, 112
Gallegos, Beatrice, 139
Gamio, Manuel, 11, 62. *See also The Life Story of the Mexican Immigrant*
García, Alma Araiza, 46, 64
Garcia, Alma, 121
García, Fernando, 42
García, Mario, 36, 40, 90
Garcílazo, Jeffrey, 10, 22
Garris, Dr. Roy, 28
Gary, Ind., 29, 33

Garza, Carmen Lomas, 83, 150
Gaspar de Alba, Alicia, xv, 121
Gastelum, Maria del Carmen Trejo de, 26
Girl Scouts, 39, 46–47
Glenn, Evelyn Nakano, 19–20
Glenn, Susan, 52, 70
Gómez Palacios, 3
Gómez, Ana Nieto, 107, 111–12, 114. *See also* Chicana feminism and *Encuentro Femenil*
Gómez, Estela, 145–46
Gómez, Mira, 39
Gómez-Quiñonez, Juan, 117
Gonzales, Corky, 104
Gonzales, Felipe, 137
González, Deena, xv
González, Elsie, 8
González, Gilbert, 16–17, 22, 34. *See also Labor and Community*
González, María C., 121
Great Depression, 20, 22, 29, 38, 76–77, 96, 149
Green, Rayna, 39
Guadalupe I. Calvo, Chihuahua, 8
Guanajuato, Mexico, 11
Guerin-Gonzales, Camille, 9
Guillen, Father Patricio. *See Calpulli*
Gutiérrez, David, 6, 87, 90, 117, 119. *See also Walls and Mirrors*
Gutiérrez, Juana, 142–43
Gutiérrez, Ramón, xv, 65, 109

Haas, Lisbeth, xv
Habermas, Jürgen, 35, 49
hair bobbing, 55
Hancock, Velia García, 118–19
Harrison, Gloria Macias, 107–8. *See also El Chicano*
Healey, Dorothy Ray, 72, 77; Cal-San organizer, 81–82; Communist Party, identification with, 74. *See also* United Cannery, Agricultural, Packing, and Allied Workers of America (UCAPAWA)
Heinze, Andrew, 52, 70
Hermosillo, Mexico, 62
Hernández, Isabel, 116
Hernández, María, 102, 115. *See also mutualistas*
Hernández, Antonia, 113, 124
Hershatter, Gail, 128, 130, 132
Hijas de Cuauhtemoc, 108
Hispana, meaning of term, xiv
Hispanic, meaning of term, xiv

Hispano America, 57
Hispanos, 6
Holding the Line. See Kingsolver, Barbara
Holguin, Margaret, 42
Hollywood, Calif., 14, 57–58, 85, 96
Hom, Alice, 122
Honig, Emily, 128, 130, 132
Hotel and Restaurant Employees Union, Local 11 (Los Angeles). *See* María Elena Durazo
Houchen Community Center. *See* Rose Gregory Houchen Settlement House
Houchen, Rose Gregory, 41. *See also* Rose Gregory Houchen Settlement House
Houston, Texas, 108
Hoyt Street, 66, 138. *See also* Ponce, Mary Helen
Huerta, Dolores, 119, 134, 143. *See also* United Farm Workers
Hughes, Howard, 85
Hull House, 34
Hurtado, Albert, xv

Ibañez, Carlos Velez, 113
Ibarra, Estella, 44–45
Idar, Jovita, 99. *See also* El Primer Congreso Mexicanista
Immigration Act of 1917, 13
Immigration and Naturalization Service (INS), 12, 20–21, 149; Boards of Special Inquiry, 11–13; Revueltas, Rosaura, harassment of, 85. *See also Salt of the Earth*
Infinite Divisions, 100
International Ladies Garment Workers Union (ILGWU), 118
International Union of Mine, Mill and Smelter Workers, 85. *See also Salt of the Earth*

Jackson, Donald, 85
Jalisco, Mexico, 4, 11
Japanese Americans, 75–77; internment of, 84; Nisei, 54. *See also* El Monte Berry Strike of 1933
Jaramilo, Cleofas, xiii
Jencks, Clint and Virginia, 85. *See also Salt of the Earth*
John Steinbeck Committee to Aid Agricultural Organization, 78
Justice for Janitors, 137
juvenile delinquency, 83–84
Juventud Católica Feminina Mexicana (JCFM), 66

Kleven, Victor, 94. See also Ortiz y Pino, Concha
Kern County, Calif., 75
Kerr, Louise Año Nuevo, 50
Kettleman City, Calif., 144
Kingsolver, Barbara, 135–36
kissing, 67

La Confederación de Uniones de Campesinos y Obreros del Estado de California (CUCOM), 76
La Conferencia de Mujeres Por La Raza, 108; controversy over feminism at, 108–9
la dueña, 51, 63, 65, 70. See also familial protectiveness
La Malinche; origins of, 106; "The Curse of La Malinche," 106; Corpi, Lucha, "Marina Mother," 107; Chicana feminist, as, 107
la migra, 29, 32. See also Immigration and Naturalization Service
La Mujer Moderna, 99. See also Villarreal, Andrea and Teresa
"La Nueva Chicana," 100, 156. See also Correa, Viola
La Opinión (Los Angeles), 56–58, 67, 83.
La Raza Unida, 100, 102, 105, 112; Mujeres por la Raza, 116; political origins of, 115–16
Labor and Community, 16. See also González, Gilbert
Lamphere, Louise, 137
Las Cruces, N. Mex., 127
"Las Pelonas," 55
Las Posadas, 26
Latina, meaning of term, xiv
League of United Latin American Citizens, (LULAC), 31, 89–90, 97; Houchen Community Center, at, 49; Montemayor, Alicia Dickerson, 73, 91; Mendez v. Westminster, 90; Pecan Sheller's Strike of 1938, 90; women's activities in, 91. See also NAACP
Legion of Decency, 66. See also Catholic Church
León, Livia, 67
Levi-Strauss, 149
Levya, Yolanda Chávez, 122
Linares, Clemente, 17–18, 20
Lipsitz, George, 52, 58. See also Time Passages
Little, Dorothy, 37, 44

Longauex y Vasquez, Enriqueta, 108
López, Mary Lou, 47
López, Sonia, 108
López, Yolanda, 11, 150
Los Angeles Community College District, 124. See also Romero, Gloria
Los Angeles County Board of Supervisors, 124. See also Molina, Gloria
Los Angeles Dressmakers Strike (1933), 131
Los Angeles Police Department (LAPD); high school student strike, during, 104
Los Angeles Times, xvi, 104; Chicano Moratorium, 115; on sterilization of Mexican-American women, 114
Los Angeles, Calif., 14, 33, 59, 103–4, 113–15, 137, 150–51; Mexican population of, 6; forced repatriation of Mexicans, 29
Los Ganados del Valle, 145
Loyola University, Los Angeles, 103
Lucas, María Elena, 133–34
Lucero, Lucy, 42
Lucero-Trujillo, Marcela Christine, 120
Lucia, R., 8
Luna, Grace, 17
Luna, Mary, 53, 61

Madera Cotton Strike, 78
Madero uprising, 8
Madrigal v. Quilligan, 113–14. See also sterilization
Magón, Juan Flores, 96
Making Face, Making Soul/Haciendo Caras, 121–22, 125. See also Anzaldúa, Gloria.
Margolis, Eric, 17
Márquez, Benjamin, 90
Martin, Carlotta Silvas, 26
Martin, Patricia Preciado, 24. See also Songs My Mother Sang to Me
Martínez, Elizabeth, 100, 105, 107, 149. See also El Grito
Martínez, Sally, 114
Marx Ferree, Myra, 101
Matsumoto, Valerie, 53
Matute-Bianchi, María Eugenia, 120
Max Factor cosmetics, 54
McCormick, Paul, 84. See also Mendez v. Westminster
McCrea and Son, strike at, 136
McDonnell-Douglas, 68, 101–2
McLean, Reverend Robert, 41

McWilliams, Carey, 29, 72, 83
Medina, Lara, 142
Mediola, Tatcho, 116
"Melting Pot," 37
Mendez v. Westminster, 84
Merino, Yamira, 150
Mestiza; feminist understanding of, 125;
  meaning of term, xiv
Methodist Church, 33–34, 37. *See also*
  Rose Gregory Houchen Settlement
  House
Mexicali, Mexico, 96
Mexican American Legal Defense Fund
  (MALDEF), 124. *See also* Hernández,
  Antonia
Mexican Chamber of Commerce, 55
Mexican Consul, 87–88
Mexican immigration, assimilation, 25;
  destinations, 9–10; occupations, 9–10;
  "push/pull" model, 7, 10; reaction
  against, 28–30, 147
Mexican Mothers Club of the University of
  Chicago Settlement, 50
Mexican Revolution, 8
Mexican-American; meaning of term, xiv;
  per-capita income of, 98; population,
  147
Mexicana, meaning of term, xiv
"Mi Amor," 109
Michoacán, Mexico, 11
Milligan, Adele Hernández, 59
Mohanty, Chandra Talpade, xv
Molano, Elvira, 85. *See also Salt of the
  Earth*
Molina, Gloria, 124, 143. *See also* Los
  Angeles County Board of Supervisors
Molina, Priscilla, 39
Moncaya, Hortensia, 99. *See also* El
  Primer Congreso Mexicanista
Montemayor, Alicia Dickerson. *See*
  LULAC
Montiel, Fermín, 66–67
Mora, Magdalena, 117–18. *See also* CASA
Mora, Pat, 119, 150, 157–58
Moraga, Cherríe, 110, 121, 150. *See also
  This Bridge Called My Back*
Morales, Elisa "Elsie," 62
Moreno, Graciela Martínez, 133
Moreno, Luisa, 73, 100, 151; American
  Committee for the Protection of the
  Foreign Born (ACPFB), 97; departs for
  Guatemala, 84; Cal San organizer,
  81–82, 101; Communist Party member-
  ship, 84; El Congreso organizer, 94–95;

El Nogal Union, 79–80; family back-
  ground, 96. *See also* UCAPAWA
Moreno, María, 47
Mothers of East Los Angeles, 142–43
Mount, George, 102. *See also* Mount, Julia
  Luna
Mount, Julia Luna, 15, 64, 82–83, 115;
  biography of, 101–2
Mount, Tanya Luna, 115
Mujeres Activas En Letras Y Cambio
  Social (MALCS), 122–23; "Animal
  Attraction," controversy over, 124
Mujeres Mexicanas, 144
Mulligan, Rose Escheverría, 45, 67
Musquíz, Virginia, 115
mutualistas, 55, 86–89, 102; Alianza
  Hispano-Americana, 86; Club Femenino
  Orquidia, 86–87; Club Mexicano
  Independencia (CMI), 86; el Comité de
  Vecinos de Lemon Grove, 88–89; La
  Asociación Hispano-Americana, 140; La
  Cruz Azul, 88; Liga Protectora Latina,
  88; Orden Cabelleros de America, 102;
  Sociedad Josefa Ortiz de Dominguez,
  87; Sociedad Mutualista Hispano-
  Azteca, 88

National Association for the Advancement
  of Colored People (NAACP), 90. *See
  also* LULAC
National Chicano Youth Liberation
  Conference, 104; controversy over femi-
  nism, 108
National Industrial Recovery Act (NIRA),
  76
National Labor Relations Board (NLRB),
  129, 136–37
National Women's Party, 91
Nava, Yolanda, 114
Negrete, Theresa, 84
New York, N.Y., 91
Newark Methodist Maternity Hospital, 36.
  See also Rose Gregory Houchen
  Settlement House
Newark-Houchen News, 45
Nicholas, Elizabeth, 74
No Separate Refuge, 45. *See also* Deutsch,
  Sarah
Nogales, Ariz., 12
Nu-Pike amusement park, Long Beach, 59
nuclear freeze movement, 102. *See also*
  Mount, Julia Luna, 102
Nueces County, Texas, 9
Nyssa, Oregon, 10

Ochoa, Kathy Ledesma, 118. *See also* CASA
"Okie" migration, 30, 77. *See also* Dust Bowl
Old Spain and the Southwest. See Warren, Adelina Otero
Olguin, Carmen, 114
Olsen, Culbert, 97
Orange County, California; citrus communities of, 16, 22
Orellana, Trinidad, 12. *See also* Immigration and Naturalization Service
Organizing for Our Lives: New Voices from Rural Communities. See Street, Richard
Orozco, Cynthia; on Chicana feminism, 125–26; 73 74, 90, 102
Ortiz y Pino, Concha, 73, 91–93; bilingual education, 94; marriage to Victor Kleven, 94; the New Deal, views on, 93–94
Ortiz, Cástulo, 87
Ospinaldo, Orinio, 141
Otero, Miguel, 92
Our Lady of Guadalupe Shrine, 22, 66

pachucas, las, 83
pachucos, 53
Pacific Electric Railway Company, 74
Padilla, Polita, 45
Palacios, Monica, xiv
Pardo, Mary, 142–43
Parra, Edna, 39
Partido Liberal Mexicano, 96
Pascoe, Peggy, 47
pateras (midwives), 25
Paul, Alice, 91
Payan, María, 47
Peace and Freedom Party, 102
Peña, Soledad, 99. *See also* El Primer Congreso Mexicanista
Pérez de Valdez, Jacinta, 60
Pérez, Emma, 122
Pérez, Selena Quintanilla ("Selena"), 150
Pesotta, Rose, 131
Pesquera, Beatríz, 125. *See also Building With Our Hands*
Phelps Dodge copper strike (1983), 135–36
Phoenix, Ariz., 104
Pocatello, Idaho, 21
Ponce, Mary Helen, 66. *See also Hoyt Street*
Progressive Era, 34
Proposition 187, 147, 151

Public space, concept of claiming, xiii, xv–xvi, 49, 73, 127–28, 145

Quevedo, Eduardo, 73, 95
Quinn, Anthony, 96
Quiñonez, Naomi, 107, 150–51

railroad work, 11, 22–24
Rainbow Room, 69
Ramírez, Mariana, 85
Ramírez, Sara Estela, 99, 102
Rebolledo, Diana, 92
Relations of Rescue. See Pascoe, Peggy
Resurrection Parish, 142
Revueltas, Rosaura, 85. *See also Salt of the Earth*
Rickford, Millie, 36–38, 42
Rico, María, 47
Rillito, Ariz., 67
Riordan, Richard Mayor, 141
Rivera, Aida, 39
Rivera, Carmen, 85. *See also Salt of the Earth*
Riverside Ballroom, 60
Riverside, Calif., 7, 22
Robert Kennedy, 104
Roberts, Kenneth, 28
Robles de Mendoza, Margarita, 86–87
Rock Island Railroad, 10, 24
Rockefeller, Nelson, 130
Rodríguez, Gregorita, 61
Rodríguez, Raymond, 11
Rodríguez, Severiano, 9
Romero, Gloria, 123–24
Romero, Mary, 17
Romo, Ricardo, 9
Roosevelt High School, Los Angeles, 104
Roosevelt, Franklin Delano, 76
Rose Gregory Houchen Settlement House, 32, 33–34; "Americanization" activities of, 36, 46; bilingual education, 40; constitution of, 48; "Friendship Square," 42–43, 45, 46–48; goals of, 35–36; Latina missionaries, 47; named for, 37; Protestantism of, 43–45; use of by Mexican-American women, 48–50
Rose, Margaret, 134
Rosie the Riveter, 82
Roybal, Hon. Edward, 130
Ruiz, Alice, 39
Ruiz, Erminia, 27, 62, 68–69
Ruiz, Maria de las Nieves Moya de, 69, 71

Ruiz, Petra, 112
Ruiz, Vicki, 11, 132; childhood of, xiii; experiences at Stanford, 121

Sal si Puedes, 68
Salas, Elizabeth, 92
Salazar, Rosalía, 24
Salazar, Ruben, 115, 119. *See also* Chicano Moratorium
Saldívar-Hull, Sonia, xv
Salt of the Earth, 84–86, 136
San Antonio, Tex., 18; Chicano student walkout, 104; Chicano activists' reunion, 121; COPS, 138; El Nogal, 79, 84; Levi-Strauss, 149; Pecan Sheller's strike of 1938, 90
San Bernardino, Calif., 6–7
San Diego County, Calif., 6
San Francisco News, 77–78
San Joaquin Valley cotton strike of 1933, 20, 74–75
San Joaquin Valley, 96, 119
San Julian, Mexico, 4, 30
Sánchez, Angela, 86. *See also Salt of the Earth*
Sánchez, George, 13, 34, 44, 95
Sánchez, Petra, 4, 30–31
Sánchez-Walker, Marjorie, 4, 30
Sandoval, Chela, 145
Santa Fe County, Tex., 92
Sarmiento, Clara, 44–45, 47
Scottsbluff, Neb., 84
segregation, 68
Segundo Barrio, 33, 44, 48; demographics of, 34–35. *See also* El Paso, Tex.
Segura, Denise, 125
Sekul, Joseph, 139
Seligman, Julius, 79
separate spheres, concept of, xvi
Shelit, Alicia Mendeola, 46, 61, 82
Sheridan, Thomas, 25, 60
Shriver, Sergeant, 130
Silvis, Ill., 10, 23
Simon J. Lubin Society, 78
Singing for My Echo, 61. *See also* Rodríguez, Gregorita
Sisters of the Company of Mary, 26
Skerry, Peter, 139–40, 143
Smith, Clara, 39, 60
Soja, Edward, 40
solas (single mothers), 11
soldaderas, 8, 111–12. *See also* Chicana feminism
Soldatenko, María, 149

Songs My Mother Sang to Me, 24–25. *See also* Martin, Patricia Preciado
Sosa Riddell, Adaljiza, 119, 123
Sosa, Gregoria, 23
Soto, Elizabeth, 42, 47
South Central Organizing Committee (SCOC), 141
Southern Pacific Railroad, 23
Southwest, as a concept, xv
Spanish American, meaning of term, xiv
Spanish-Mexican settlement; myth about, 5
Spaulding, Charles, 77
Stanford University, Calif., 120–21
Stark, Herman, 83
Steinbeck, John, 20, 77. *See also The Grapes of Wrath*
sterilization, 113–14. *See also* USC/Los Angeles County Medical Center
Stoltz, Effie, 36. *See also* Freeman Clinic
Street, Richard, 143–44. *See also Organizing for Our Lives*
Student Non-Violent Coordinating Committee (SNCC), 104, 144
Students for a Democratic Society, 103
Sunbelt Working Mothers. See Evans, Peter; Gonzales, Felipe; Lamphere, Louise; Zavella, Patricia
Superior, Ariz., 25

Tafolla, Carmen, 122–23. *See also To Split a Human*
Tarrango, Yolanda, 138
Taylor, Paul, 9, 45, 59
Teamsters Union, 84
Tejana, meaning of term, xiv
Tellez, Jesse, 136
Tenayuca, Emma, 79
Tenépal, Malintzin. *See* La Malinche
The Arizona Teacher and Home Journal, 38
The Chicana Feminist, 119. *See also* Cotera, Martha
The Grapes of Wrath, 77, 97. *See also* Steinbeck, John
The Life Story of the Mexican Immigrant, 11
The Saturday Evening Post, 28
Third World Strike at UC-Berkeley (1969), 104
This Bridge Called My Back, 110, 121–22. *See also* Moraga, Cherríe; Anzaldúa, Gloria
Thompson, E.P., 74

Time Passages, 58. *See also* Lipsitz, George, 58

To Split a Human, 122–23. *See also* Tafolla, Carmen

Torres, Jesusita, xvii, 3–4, 14–16, 18–19, 30, 52, 75–77

Treaty of Guadalupe-Hidalgo, 5

Treviño, Roberto, 56

Treviño-Sauceda, Millie, 144, 150

Triangle Shirtwaist Factory, 149

Trinidad, Colo., 27

True Story magazine, 52, 81

Trujillo, Carla, 121–22

Tuck, Ruth, 6–7, 46, 67–68

Tucson, Arizona, 60

Tulare County, California, 75

U.S. Chamber of Commerce, 9–10

U.S.-Mexican War, xv, 5;

UCLA Film and Television Department, 124

Unbound Feet. See Yung, Judy

Unidad Para Siempre. See Farah textile strike

United Auto Workers, 82, 102. *See also* Mount, George

United Cannery, Agricultural, Packing, and Allied Workers of America (UCAPAWA-CIO), 77–79, 140; cannery workers in, 80–82; gender awareness, 101; Healey, Dorothy Ray, 72, 77; Moreno, Luisa, 73, 79–80, 96; red scare, 84; walnut workers, in, 72–73

United Farm Workers (UFW), 103, 105; background of, 132; Farah textile strike, support for, 130; grape boycott, 119, 132; UFW Service Center, 133. *See also* Chávez, Cesar; Chávez, Helen; Huerta, Dolores; Lucas, María Elena

United Mexican American Students, 103

United Neighborhood Organization (UNO), 141

"University Avenue." *See* Mora, Pat

University of California at Los Angeles (UCLA), 113, 117

University of California at Santa Barbara, 105

USC/Los Angeles County Medical Center, 113–14

Utah-Idaho Sugar Company, 9, 21, 23

Valdez, Patíssi, 103

Varela, María, 104–5, 144–45

Vargas, Zaragosa, 79

Vásquez, Felix and Frederica, 10, 22–24

Velásquez, Baldemar, 133. *See also* Farm Labor Organizing Committee

Vélez, Julia Yslas, 25

Vellanoweth, Patricia, 118. *See also* CASA

vergüenza (female chastity), 65. *See also* familial protectiveness

Veyna, Angelina, xv

Victoria, Tex., 127

Vietnam War, 105, 114–15. *See also* Chicano Moratorium

Villarreal, Andrea and Teresa, 99. *See also* *La Mujer Moderna*

Walls and Mirrors, 89. See also Gutiérrez, David

Walsenberg, Colorado, 27

Warren, Adelina Otero, 91–92. *See also* *Old Spain and the Southwest*

We Fed Them Cactus, 24. *See also* Cabeza de Baca, Fabiola

Weber, Devra, 8, 11, 16, 20, 74–75. *See also Dark Sweat, White Gold*

Weinberg, Sydney Stahl, 52

Welty, Eudora, 145

Whatley, Charlotte, 91. *See also* Warren, Adelina Otero

White, Richard, 27

Women's Home Missionary Society of Newark, New Jersey, 41

Women's Work and Chicano Families. See Zavella, Patricia

Workers' Alliance, 79, 101

World War I, 9

World War II, 53, 90, 101–2

Xicana, meaning of term, xiv

Xicanisma, 125

Ybarra, María, 51, 61

Yung, Judy, 53. *See also Unbound Feet*

Zamora, Bernice, 70–71

Zarate, Sister Rosa Marta. *See also* *Calpulli*

Zavella, Patricia, 136–37. *See also Sunbelt Working Mothers* and *Women's Work and Chicano Families*

Zinn, Maxine Baca, 135

Zoot Suit Riots, 83–84